Continuous Cultures of Cells

T0231125

of

Cells

Volume II

Editor

Peter H. Calcott, D. Phil.

Assistant Professor
Department of Biological Sciences
Wright State University
Dayton, Ohio

CRC Press
Taylor & Francis Group
Boca Raton London New York

CRC Press is an imprint of the
Taylor & Francis Group, an **informa** business

First published 1981 by CRC Press
Taylor & Francis Group
6000 Broken Sound Parkway NW, Suite
300 Boca Raton, FL 33487-2742

Reissued 2018 by CRC Press

© 1981 by Taylor & Francis
CRC Press is an imprint of Taylor & Francis Group, an Informa business

No claim to original U.S. Government works

A Library of Congress record exists under LC control number: 81001656

Publisher's Note
The publisher has gone to great lengths to ensure the quality of this reprint but points out that some imperfections in the original copies may be apparent.

Disclaimer
The publisher has made every effort to trace copyright holders and welcomes correspondence from those they have been unable to contact.

ISBN 13: 978-1-138-50579-7 (hbk)
ISBN 13: 978-1-138-55826-7 (pbk)
ISBN 13: 978-1-315-15053-6 (ebk)

Visit the Taylor & Francis Web site at http://www.taylorandfrancis.com and the CRC Press Web site at http://www.crcpress.com

PREFACE

Continuous culture is a method used both in research and industry to grow microbes, primarily bacteria; though it has been used to grow algae, protozoa, fungi, and plant and animal cells. Continuous, or open, culture differs from the batch, or closed, culture method in so much as it protracts growth of the organisms in a time independent dimension. With this method, which is incidently more complex to operate than simple batch culture, it is possible to study more rigorously the physiology, biochemistry and genetics of microorganisms as it relates to, for instance, the influence of environmental factors.

Previous to this book, a number of reviews, symposia, and monographs have focused on continuous culture as a tool. However, no one book or symposium has managed to capture the full range of applications. This book was intended to attain this goal. The inception of the project was in 1977 when I invited some world-recognized experts in the field to contribute chapters to this treatise. The book was compiled by the summer of 1978. Thus this book documents and illustrates the knowledge known and recognized to that date. There has been, as in all rapidly advancing areas, an advance in this area since 1978 which has obviously not been included.

I have tried in this book to present as broad a perspective as possible to the subject matter. In the construction of the chapters, I have also left much up to the individual contributors. Some chapters have been written as essentially up to the minute reviews of an application or use of continuous culture while others have used data obtained in the author's own laboratory to illustrate the use of continuous culture as a problem-solving tool. Yet others have concentrated on specific topics and cited a few key ways in which continuous culture can be useful. The approach was left solely up to the contributor.

In this two-volume set, I have included chapters on the overall perspective of the technique (Chapter 1), the construction and operation of laboratory cultures (Chapter 2), and the mathematics of growth in continuous cultures (Chapter 3). Chapter 4 focuses on the use and advantage of complex and multi-stage continuous fermenters while Chapters 5 and 6 demonstrate the use of the technique in industry to produce single cell protein and fine biochemicals and drugs from simple or waste substrates. Chapter 7 demonstrates the use of the technique in studying nonsteady state or transient phenomena. Volume II also comprises seven chapters, with the first four being devoted to the application of the technique to the studies of cell metabolism, more specifically to carbon metabolism, chemical composition of cells, intermediary metabolism, and oxidative phosphorylation. Studies on the genetics of microorganisms in continuous culture is dealt with in Chapter 5. Plant cell and algae culture are focused in Chapters 6 and 7. While these two volumes cover most applications of the technique there are several which because of space were not included; these are discussed in Chapter 1.

It is hoped that this two-volume set will be useful to the established continuous culture operator as well as the researcher, teacher, and student who is interested in learning how the technique could be useful in answering both basic and applied questions in microbiology and cell biology.

Peter H. Calcott
January, 1981

THE EDITOR

Peter H. Calcott, D. Phil., is Assistant Professor of Biological Sciences at Wright State University, Dayton, Ohio.

Dr. Calcott was graduated from the University of East Anglia, Norwich, England, with a B.Sc. (Hons.) degree in Biological Sciences in 1969. He received his D. Phil. (Biology) in 1972 from the University of Sussex, England under the supervision of Professor J. R. Postgate. After 2 years postdoctoral training in the Department of Microbiology, Macdonald College of McGill University, Montreal, Quebec, Canada with Professor R. A. MacLeod, Dr. Calcott was a Professional Associate in that department from 1974 to 1976. He joined the faculty of Biological Sciences at Wright State University in 1976. Dr. Calcott has spent leaves working with Dr. D. Dean, Department of Microbiology, Ohio State University and Professor A. H. Rose, University of Bath, England.

Dr. Calcott's research interests revolve around the reaction of microbes to stress, primarily freezing and thawing and starvation. He is particularly interested in the role of cell wall and membrane structures in determining the resistance of organisms to stress, continuous culture of microbes and the role of small molecules such as cyclic AMP and cyclic GMP in cell metabolism. Dr. Calcott has published more than 50 research papers, reviews, books, and abstracts over his career.

CONTRIBUTORS

M. J. Bazin
Senior Lecturer in Microbiology
Microbiology Department
Queen Elizabeth College
London, England

Peter H. Calcott
Assistant Professor
Department of Biological Sciences
Wright State University
Dayton, Ohio

Sallie W. Chisholm
Associate Professor
Doherty Professor of Ocean
  Utilization
Division of Water Resources and
  Environmental Engineering
Civil Engineering Department
Massachusetts Institute of
  Technology
Cambridge, Massachusetts

F. Constabel
Senior Research Officer
Prairie Regional Laboratory
National Research Council of
  Canada
Saskatoon/Saskatchewan
Canada

Charles L. Cooney
Associate Professor of Biochemical
  Engineering
Biochemical Engineering Laboratory
Department of Nutrition and Food
  Science
Massachusetts Institute of
  Technology
Cambridge, Massachusetts

E. A. Dawes
Department Head
Reckitt Professor of Biochemistry
Department of Biochemistry
University of Hull
Hull, England

P. Dobersky
Research Specialist
Department of Technical
  Microbiology
Czechoslovak Academy of Sciences
Praha, Czechoslovakia

J. W. Drozd
Doctor
Shell Research Limited
Shell Biosciences Laboratory
Kent, England

D. C. Ellwood
Director
Pathogenic Microbes Research
  Laboratory
PHLS Centre for Applied
  Microbiology & Research
Wiltshire, England

Ivan J. Gotham
Research Scientist I
New York State Department of
  Health
Albany, New York

Margareta Häggström
Doctor
Technical Microbiology Chemical
  Center
Lund University
Lund, Sweden

Walter P. Hempfling
Associate Professor
Department of Biology
The University of Rochester
Rochester, New York

H. Michael Koplov
Section Leader
Schering Corporation
Union, New Jersey

W. G. W. Kurz
Senior Research Officer
Prairie Regional Laboratory
National Research Council of
    Canada
Saskatoon/Saskatchewan
Canada

J. D. Linton
Shell Research Limited
Shell Bioscience Laboratory
Kent, England

Abdul Matin
Assistant Professor
Department of Medical
    Microbiology
Stanford University
Stanford, California

G-Yull Rhee
Research Scientist IV
New York State Department of
    Health
Albany, New York
Adjunct Associate Professor
Cornell University
Ithaca, New York

Craig W. Rice
Fellow
Department of Biochemistry and
    Biophysics
Division of Genetics
University of California San
    Francisco
San Francisco, California

J. Řičica
Senior Scientific Worker
Deputy Head
Department of Technical
    Microbiology
Institute of Microbiology
Czechoslovak Academy of Sciences
Praha, Czechoslovakia

A. Robinson
Head of Pertussis Vaccine Unit
PHLS Centre for Applied
    Microbiology & Research
Pathogenic Microbes Research
    Laboratory
Wiltshire, England

V. R. Srinivasan
Professor
Department of Microbiology
Louisiana State University
Baton Rouge, Louisiana

R. J. Summers
Senior Research Biologist
M. E. Pruitt Research Center
Dow Chemical, U.S.A.
Midland, Michigan

TABLE OF CONTENTS

Volume I

Volume II

Chapter 1

CARBON METABOLISM*

Edwin A. Dawes

TABLE OF CONTENTS

* This chapter was submitted in Febuary 1978.

I. INTRODUCTION

A. Objectives

This contribution has been designed to survey the distinctive advantages conferred by the application of the continuous culture technique to studies of carbon metabolism, to point to some of the essential features of methodology, and to describe its application to two specific problems in bacterial carbon metabolism. There has been no attempt to review comprehensively all those investigations of microbial metabolism of carbon compounds that have utilized continuous culture, and the examples recorded in detail are taken principally from researches carried out in our own laboratory. These are, first, the investigation of the regulation of carbohydrate transport and metabolism in *Pseudomonas aeruginosa* and, second, the role and regulation of the carbon and energy reserve polymer, poly-β-hydroxybutyrate, in *Azotobacter beijerinckii*. These projects afford some insight into the study of competing carbon substrates and of the influence of different growth limitations on carbon metabolism, and they illustrate also the application of gaseous limitations. For other examples of mixed substrate utilization in chemostats the reader is referred to the review by Harder and Dijkhuizen. [1]

B. The Value of Continuous Cultivation Techniques

The pathways of carbon metabolism in bacteria can be profoundly affected by the environment, as manifest by the availability and type of carbon and nitrogen sources and inorganic ions, by pH, and by the partial pressure of oxygen. In order to investigate systematically metabolic pathways and their regulation, it becomes imperative, therefore, that the experimenter should be able to control the environment so that the effects of these various individual variables can, in turn, be investigated. The advantages of continuous culture for such studies are patent; this technique enables the bacterial culture to be held in a steady state in a defined environment at a growth rate determined by a single nutrient. Usually this limiting nutrient is supplied in the inflowing medium at a rate determined by the dilution rate, but in the case of gaseous limitations such as oxygen, or nitrogen in the case of nitrogen-fixing organisms, the flow rate of the gas into the culture is the controlling factor and parameters such as the solubility coefficient of the gas in the medium and gas transfer coefficient must be taken into account.

The chemostat permits not only the investigation of enzyme levels and metabolism in steady-state systems, but also the important changes that occur during the transient states that exist between steady states in response to qualitative or quantitative alterations in the factors limiting growth.

II. METHODOLOGY

In the studies of carbon metabolism, choice of the most suitable chemostat is important. The working volume requires careful consideration in relation to the size of samles that need to be taken for analysis. The steady state will be drastically upset if a significant proportion of the culture is removed. Clearly, the higher the organism concentration in the culture the smaller the volume of sample taken need be, but operational problems can arise with too dense cultures in ensuring that unwanted gas limitations do not occur, that wall growth does not become a serious problem, and that foaming is minimized. The supply of antifoaming agents to the culture must be kept to the minimum rate compatible with the desired objective, since excess can be taken up by bacteria and, even after thorough washing, be carried through into the cell extracts where interference with spectrophotometric assays may occur. The general observation that prevention is better than cure applies especially to foaming in chemostat cultures, and thus a little time spent in ascertaining the optimal supply rate to avoid foaming and troublesome carry-over of antifoam agent pays subsequent dividends.

Automatic control of pH is an essential requirement for investigation of carbon metabolism because of the profound effect of pH on metabolic pathways. In our opinion the use of heavily buffered media is not a satisfactory method of control because of the undesirable osmoregulatory effects imposed on the organism by a high salt concentration. If large quantities of acid are produced or utilized in metabolism, then the volume of compensating alkali or acid added to the culture per hour can represent a significant proportion of the total culture volume, and appropriate corrections should always be made.

In studies of the effect of varying growth (dilution) rate on metabolism, the consequential effects on the composition of the organism must be borne in mind since the ribonucleic acid content is a function of the growth rate (Table 1), and also significant differences may occur in the specific activities of enzymes (see Table 2) as a consequence of changing intracellular concentrations of inducers and/ or repressors in response to altered growth rate. Further, change of dilution rate requires attention to the rate of gas supply to the culture to ensure that inadvertant limitation, e.g., of oxygen, does not occur.

III. CARBOHYDRATE METABOLISM IN *PSEUDOMONAS AERUGINOSA*

A. Glucose Metabolism and Diauxic Growth

Glucose metabolism in *Pseudomonas aeruginosa* is complex and the investigations of various research groups[2-6] revealed the apparent existence of the pathways shown in Figure 1. Glucose may be oxidized to gluconate and 2-oxogluconate prior to phosphorylation, or may be phosphorylated first and then oxidized. The discovery of a kinase for 2-oxogluconate and a reductase for 2-oxogluconate 6-phosphate[6] meant that all these pathways converged on gluconate 6-phosphate, which was then further metabolized principally via the Entner-Doudoroff route and, according to radiorespirometric experiments, to a minor extent via the pentose phosphate cycle[5]. Recent studies with mutants devoid of gluconate 6-phosphate dehydratase, and thus deficient in the Entner-Doudoroff pathway, showed, however, that such organisms were unable to catabolize glucose and that gluconate 6-phosphate dehydrogenase, furnishing entry to the pentose cycle, was therefore not operative[7]. There is not a functional glycolytic system, since the organism lacks phosphofructokinase,[8,9] although growth occurs readily on organic acids and gluconeogenesis operates via fructose 1,6—bisphosphate aldolase with pentose formation via transketolase and transaldolase carbon rearrangements.[9] Terminal oxidation occurs via the tricarboxylic acid cycle, which is constitutive in *P. aeruginosa*, although the entry of various intermediates of the cycle into glucose-grown bacteria is mediated by inducible permeases.[2,10]

When *P. aeruginosa* was grown in a mineral salts medium with citrate or succinate as the sole carbon source, or in peptone medium, the principal enzymes of glucose metabolism were severely repressed.[11-13] However, growth on glycerol led to derepression. Batch culture in media containing glucose and an organic acid resulted in diauxic growth with the organic acid being the preferential substrate.[14] Hamilton and Dawes[11,12] examined three possible explanations for the observed diauxie, namely: (1) the enzymes of glucose metabolism are constitutive, but entry of glucose is mediated by a transport system that is repressed by growth on organic acids, (2) some or all of the glucose-catabolizing enzymes are inducible and their formation and/or activities

Table 1
EFFECT OF DILUTION RATE ON RNA AND NITROGEN CONTENT OF *P. AERUGINOSA* EXTRACTS

Dilation rate	Concentration in bacterial extract (mg ml^{-1})				
(hr^{-1})	0.125	0.170	0.200	0.250	0.500
Total N	2.70	2.77	2.81	2.75	2.84
RNA	1.60	1.96	2.16	2.43	2.76
RNA N	0.26	0.31	0.35	0.38	0.44
RNA N/total N (%)	9.6	11.2	12.5	13.8	15.5

From Ng, F. M-W. and Dawes, E. A., *Biochem J.*,132, 141, 1973. With permission.

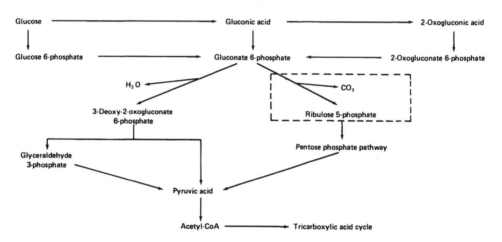

FIGURE 1.   Pathways of glucose metabolism in *Pseudomonas aeruginosa* as originally envisaged. Subsequent work revealed that gluconate 6-phosphate dehydrogenase is not present so that ribulose 5-phosphate cannot be formed oxidatively.

are repressed or inhibited by the simultaneous presence of an organic acid, but the glucose transport system is constitutive, or (3) both the glucose enzymes and the transport system are inducible. Evidence was obtained for the presence of an inducible transport system for glucose, and organisms grown on succinate, citrate, or peptone had very low activities of the glucose-catabolizing enzymes; incubation of washed suspensions with glucose produced a significant increase in the activities of these enzymes, which did not occur in the presence of chloramphenicol.

Because of the inherent difficulties of investigating enzyme levels during the first phase of diauxic growth in batch culture, when bacterial densities are very low, the chemostat offers an excellent experimental approach to such studies of enzyme regulation and, moreover, provides a constant environment that cannot be attained in batch culture. Consequently the organism was grown in a chemostat and the effect of varying the relative concentrations of glucose and citrate on the key enzymes of glucose catabolism and the tricarboxylic acid cycle was examined for both steady and transient states.[13,15] Conditions of nitrogen (ammonium) limitation were chosen because it was believed that maximum catabolite repression would be observed in these circumstances.

Table 2

EFFECT OF DILUTION RATE ON ENZYMIC ACTIVITIES OF *PSEUDOMONAS AERUGINOSA* GROWN IN A CITRATE-GLUCOSE MEDIUM

Specific activity [μmol hr^{-1} (mg of N)$^{-1}$]

Dilution rate (hr^{-1})	Glucose 6-phosphate dehydrogenase	Hexokinase	Gluconokinase	Entner-Doudoroff enzymes	Glucose dehydrogenase	Gluconate dehydrogenase	Isocitrate dehydrogenase	Aconitase
0.125	35.6	14.2	3.7	3.6	0.7	21.0	449.2	79.0
0.170	33.5	15.2	3.0	4.3	2.0	30.6	451.7	65.5
0.200	27.4	19.3	1.5	5.8	4.0	32.0	378.8	51.9
0.250	24.9	17.1	0.2	6.0	1.8	34.1	322.7	30.0
0.500	62.0	25.0	2.9	5.4	4.3	37.4	321.0	21.6

Note: The inflowing glucose concentration was 4 m*M* at all dilution rates, but the citrate concentration in the inflowing medium was adjusted to secure a residual concentration in the chemostat vessel of 27 to 30 m*M* under all conditions. The values recorded are for the steady states established at each dilution rate.

From Ng, F. M-W. and Dawes, E. A., *Biochem. J.*, 132, 129, 1973. With permission.

B. The Effect of Citrate on Enzymes of Glucose Catabolism

Organisms were grown at a dilution rate of 0.25 hr^{-1} at pH 7.1 with 75 mM-citrate as the carbon source and then gradually increasing concentrations of glucose were introduced and the specific activities of the various enzymes assayed for each steady state (Figure 2). Below concentrations of 6 to 8 mM, glucose elicited little change in the specific activities of these enzymes, but above these concentrations the levels increased, most markedly with glucose 6-phosphate dehydrogenase, and further increases were manifest when the citrate concentration in the inflowing medium was decreased to 60 mM and then to 45 mM. The utilization of glucose by the culture reflected the changes in specific activity observed.

Investigation of the transient periods following the change of glucose concentration revealed that the increases were immediate and continued for above two doubling times. Figure 3 illustrates the transition from 6 to 8 mM glucose, the concentration range where the most marked inductive effect of glucose was observed.

The converse experiment of increasing the citrate concentration in 45 mM-glucose medium (Figure 4) showed a rapid induction of the citrate transport system (glucose-grown organisms possess a fully operative tricarboxylic acid cycle) and the maximum response was invoked by 8 mM- citrate. Above this concentration of citrate repression of the glucose-catabolizing enzymes occurred, the magnitude of the repression increasing with increasing citrate concentration.

A series of experiments of this type with varying concentration ratios of citrate and glucose clearly demonstrated that the specific activities of the glucose enzymes could be increased either by increasing the glucose concentration or decreasing the citrate concentration in the medium, observations which accorded with the regulation of the glucose enzymes by induction with glucose or its metabolites and repression by citrate or its metabolites.

C. Glucose Transport in *P. aeruginosa*

The discovery of a threshold glucose concentration of 6 to 8 mM for significant induction of the glucose enzymes suggested the possibility that induction of a glucose transport system was playing an important role in the phenomenon and, since transport can be a major regulatory process, attention was turned to carbohydrate transport in *P. aeruginosa*, specifically posing the questions:

1.   Is the glucose transport system repressed in citrate-grown cells?
2.   Is the *activity* of the transport system subject to metabolic regulation?

These investigations were not carried out with chemostat-grown organisms and are not, therefore, described in detail, but the findings [6] are essential for the subsequent chemostat studies and may be summarized as follows.

P. aeruginosa is sensitive to cold-shock and the standard technique for transport studies, involving washing of the organisms at 0°C, leads to erroneous results. However, the incorporation of 1% (w/v) NaCl in the washing fluid [67 mM-NaK phosphate, pH 7.1; 0.1% w/v MgSO$_4$, 7H$_2$O] and conducting the operations at 21°C, revealed that the uptake of methyl α-glucoside obeyed saturation kinetics, was energy dependent, and yielded free methyl α-glucoside as the intracellular product. The uptake process was competitively inhibited by 2-deoxyglucose, glucosamine, mannose, galactose, 6-deoxyglucose, xylose, fucose, and methy β-galactoside, indicating that changes of configuration at hexose carbon atoms 1,2,4, and 6 can be tolerated by this system.

Metabolic regulation of the methyl α-glucoside transport system was demonstrated by preincubating organisms with metabolizable substrates for short periods prior to assaying for transpor'; acetate, succinate, pyruvate, and gluconate all inhibited the

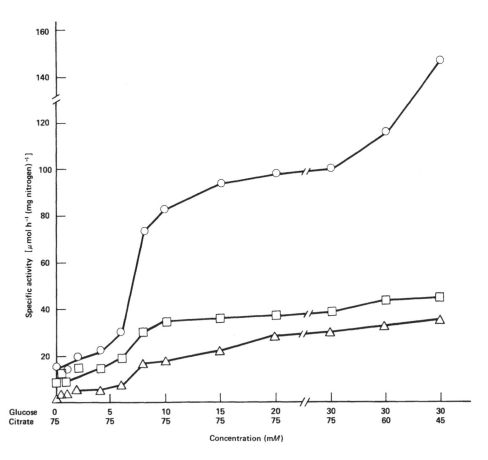

FIGURE 2. Effect of relative citrate:glucose concentrations in the inflowing medium on the steady-state enzymic specific activities of citrate-grown *Pseudomonas aeruginosa*. D = 0.25 hr⁻¹, pH 7.1. Glucose 6-phosphate dehydrogenase, O; hexokinase, □; Entner-Doudoroff enzymes, △.

uptake system. However, when a direct comparison was made of methyl α-glucoside uptake and of glucose uptake by *P. aeruginosa* grown on different carbon sources, surprisingly, an imperfect correlation was discovered (Table 3) and this led to experiments with labeled glucose as the substrate. Measurements under all the experimental conditions used showed that the initial rate of [U-¹⁴ C] glucose uptake was linear, and passed through an experimentally defined zero-time origin that was obtained with organisms that had been preincubated for 10 min with 25 mM-formaldehyde to eliminate uptake. The involvement of two components in glucose uptake was revealed (Figure 5), one a high affinity system with a K_m of 8 μM, and the other a low affinity system with an approximate K_m of 2 mM. When smilar experiments were carried out with gluconate- or glycerol-grown organisms, no uptake by the low-K_m component could be discerned, but uptake was mediated by a process with a K_m of approximately 1 mM in both types of organism.

It was then apparent that *P. aeruginosa* could accumulate glucose via two independent uptake systems. One of these was an active transport system of broad specificity, characterized as the methyl α-glucoside transport system and present only in the glucose-grown organism, and probably identical with the low-K_m glucose uptake system. The nature of the second system was next investigated, and since there was reason to believe it was in some way associated with glucose dehydrogenase activity, this enzyme was studied.

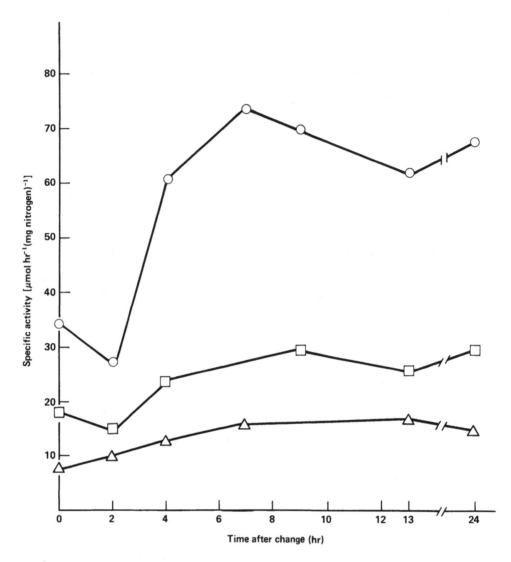

FIGURE 3. Changes in enzymic specific activities of *Pseudomonas aeruginosa* during the transient period in the chemostat after increase of glucose concentration from 6 to 8 mM in the presence of 75 mM-citrate. $D = 0.25$ hr⁻¹, pH 7.1. Glucose 6-phosphate dehydrogenase, O; hexokinase, □; Entner-Doudoroff enzymes, Δ.

Glucose dehydrogenase is a membrane-bound enzyme found in glucose-, gluconate- and glycerol-grown *P. aeruginosa*. Membrane preparations derived from glucose-grown cells displayed a K_m for glucose oxidation of 1 mM. The correlation between the K_m for glucose oxidation and the high-K_m component in glucose-, gluconate-, and glycerol-grown organisms, together with the presence of glucose dehydrogenase in such cells, suggested that the entry of glucose by the high K_m component involved in some way the activity of this enzyme. To establish this point, two, independently isolated, glucose dehydrogenase-negative mutants were examined and found to possess only the low-K_m uptake system, which was present in glucose-grown, but absent from gluconate- or glycerol-grown organisms. The high-K_m system was absent under all growth conditions, which thus indicated a requirement for glucose dehydrogenase activity for the high-K_m uptake system to be operative.

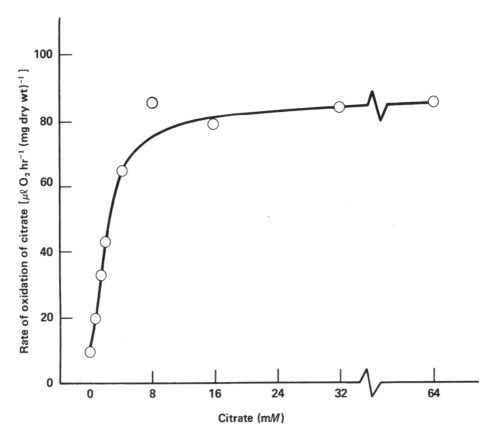

FIGURE 4. Effect of increasing the citrate concentration in the inflowing 45 mM-glucose chemostat medium on the induction of citrate permease in *Pseudomonas aeruginosa*. Chloramphenicol was present in the Warburg flasks.

Table 3
OCCURRENCE OF TRANSPORT
SYSTEMS IN *P. AERUGINOSA* 2F32
GROWN ON VARIOUS CARBON
SOURCES

Carbon source	Methyl α-gluoside transport	Glucose transport
Glucose	+	+
Gluconate	−	+
Glycerol	−	+
Citrate	−	−

From Midgley, M. and Dawes, E. A., *Biochem. J.*, 132, 141, 1973. With permission.

This finding suggested the possibility that the membrane- bound glucose dehydrogenase might be catalyzing a group translocation of glucose and releasing gluconate to the interior of the cell. However, such a mechanism was eliminated by the discovery that the rate of [U-^{14}C]glucose utilization greatly exceeded the rate of ^{14}C uptake (Table 4). These results indicate that glucose dehydrogenase acts extracellularly, converting

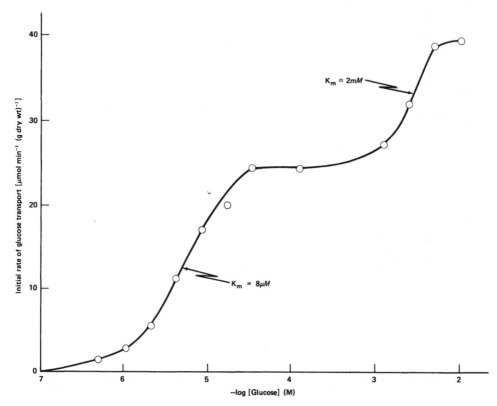

FIGURE 5.   Effect of glucose concentration on the uptake of ^{14}C from [U-^{14}C]-glucose by glucose-grown *Pseudomonas aeruginosa*.

Table 4

COMPARISON OF RATES OF UTILIZATION, AND OF
^{14}C-UPTAKE OF [U-^{14}C]GLUCOSE,
[U-^{14}C]GLUCOSAMINE AND [U-^{14}C]GLUCONATE BY
GLUCONATE-GROWN *P. AERUGINOSA*

| | | | Rate [μmol min^{-1} (g dry wt)$^{-1}$] | |
Organism	Substrate	Concentration mM	Accumulation	Utilization
2F32	Glucose	5	23	285
	Glucosamine	20	1.2	180
PAO1	Gluconate	5	30	440

Note: Strains 2F32 and PAO1 possess essentially similar biochemical properties.

From Roberts, B. K., Midgley M., and Dawes, E. A., *J. Gen. Microbiol.*, 78,
319, 1973, and Midgley, M. and Dawes, E. A., *Biochem. J.*, 132, 141, 1973.
With permission.

glucose to gluconate, which is then taken up by the cells. This interpretation was sub-
stantiated by the oberservation that the uptake of [^{14}C]glucose by gluconate- or glycerol-grown organisms was completely inhibited by the addition of 10 mM-gluconate
10 sec prior to the glucose. In contrast, the addition of 2-oxogluconate had little effect,
indicating that 2-oxogluconate, produced by the further oxidation of gluconate, was
not the principal uptake substrate under these conditions.

The significance of gluconate dehydrogenase in glucose metabolism became apparent with our discovery[6] that, contrary to a previous report,[17] the enzyme was membrane-associated, and further, that [^{14}C]gluconate was utilized at 15-fold the rate that ^{14}C entered the cells (Table 4). It was concluded, therefore, that the enzymes of the direct oxidative pathway, glucose dehydrogenase and gluconate dehydrogenase, are oriented in the membrane in such a manner that they effect the oxidation of their respective substrates extracellularly, i.e., in the periplasmic space, and the products are then taken up by specific transport systems, the nature of which we also studied.[6] The apparent complexity of glucose metabolism in *P. aeruginosa* can thus be explained on the basis that the direct oxidative pathway and the phosphorylative pathway occur in different compartments of the cell, namely extracellularly (periplasmically) and intracellularly, respectively, and are connected via inducible transport systems (Figure 6).

An examination of the uptake of gluconate and 2-oxogluconate by *P. aeruginosa* grown on various carbon sources (Table 5) revealed that glycerol-grown cells possessed the gluconate transport system, but not that for 2-oxogluconate, which also correlated with the observation that the enzymes of 2-oxogluconate metabolism, the kinase and 2-oxogluconate 6-phosphase reductase, were not detectable in such cells.[6] The transport systems for glucose, gluconate, and 2-oxogluconate were all repressed in citrate-grown *P. aeruginosa*.

D. Chemostat Studies on the Regulation of Transport of Glucose, Gluconate, and 2-Oxogluconate

The demonstration of the existence of independently regulated transport systems for glucose, gluconate, and 2-oxogluconate led next to a study of their regulation by citrate in chemostat experiments of the type previously described.[18] Ammonium-limited organisms were grown on 75 m*M*-citrate with D = 0.40 hr^{-1} and increasing concentrations of glucose were introduced. For each steady state the activities of the transport systems were measured (Figure 7). Significant induction of the glucose transport system occurred only when the glucose concentration exceeded 6 to 8 m*M* and maximum activity was recorded between 10 and 15 m*M*. Surprisingly, higher concentrations of glucose decreased the activity of glucose transport to values approaching those found for growth on 75 m*M*-citrate alone. However, these higher glucose concentrations induced the gluconate and 2-oxogluconate transport systems, the latter responding more slowly than the former. A decrease of the citrate concentration from 75 to 45 m*M* increased the activities of both these systems, (but had no effect on the glucose-transport system), while highest activities were recorded on 45 m*M*-glucose alone. Reintroduction of 75 m*M*-citrate and 4 m*M*-glucose caused virtually complete repression of these transport systems.

The effect of these various concentrations of citrate and glucose in the inflowing medium on the enzymes of the extracellular and intracellular pathways of glucose metabolism are shown in Figures 8 and 9, respectively. These enzymes of glucose catabolism showed significant induction when the glucose concentration reached 6 to 10 m*M*. Further increases in glucose concentration, or a decrease in citrate concentration, had no effect on the activities of hexokinase and glucose 6-phosphate dehydrogenase. However, higher glucose concentrations induced gluconate kinase, as well as greatly increasing the activities of glucose dehydrogenase and gluconate dehydrogenase. In a medium containing 45 m*M*-glucose alone, the enzymes of 2-oxogluconate catabolism were induced. Some maintenance of the activities of glucose and gluconate dehydrogenase was observed when the steady state was changed from 45 m*M* glucose to 75 m*M* citrate plus 4 m*M* glucose. At all glucose concentrations studied, a significant proportion of

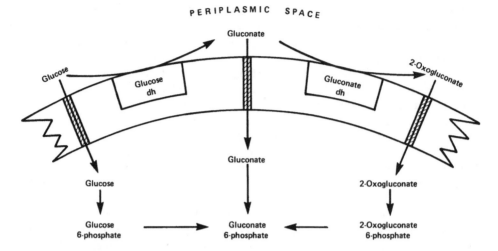

FIGURE 6.   Extracellular and intracellular metabolism of glucose, and transport systems in *Pseudomonas aeruginosa.*

Table 5

TRANSPORT OF GLUNCONATE AND
2-OXOGLUCONATE BY *P.
AETUGINOSA* PAO1 GROWN ON
VARIOUS CARBON SOURCES

Growth substrate	Uptake of [^{14}C] substrate [μmol min^{-1} (g dry wt)$^{-1}$]	
	Gluconate	2-Oxogluconate
Glucose	48	18.1
Gluconate	45	36.3
2-Oxogluconate	n.p.[a]	37.5
Glycerol	53	n.d.[b]
Citrate	n.d.[b]	n.d.[b]

[a]   n.p. = not performed
[b]   n.d. = not detected

From Roberts, B. K., Midgley, M., and Dawes, E. A., *J. Gen. Microbiol.*, 78, 319, 1973. With permission.

the residual carbon in the chemostat was present as gluconate and 2-oxogluconate, produced via the extracellular action of glucose dehydrogenase and gluconate dehydrogenase.

We had reason to suspect that the unexpected decrease in the activity of the glucose transport system at higher glucose concentrations (Figure 7) might be attributed to this gluconate, produced extracellularly by the action of glucose dehydrogenase, since gluconate-grown *P. aeruginosa* had repressed levels of the glucose-transport system and gluconate also inhibited glucose transport activity. Thus in an ammonium-limited chemostat, at a dilution rate of 0.41 hr^{-1}, the residual gluconate concentration was 2.0 to 2.2 m*M*. Consequently this possibility was examined in the chemostat by studying the interaction of citrate, glucose, and gluconate (Figure 10). Induction of the glucose uptake system followed a course similar to that of Figure 7 and repression occurred

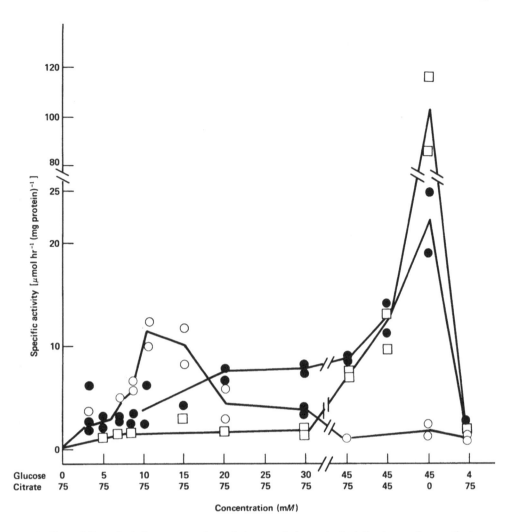

FIGURE 7. Effect of relative concentrations of citrate and glucose in the inflowing medium on the transport systems for glucose, gluconate and 2-oxogluconate in ammonium-limited *Pseudomonas aeruginosa*. D = 0.40 hr⁻¹, pH 7.1. Transport systems for glucose, O; gluconate, ●; 2-oxogluconate, □.

when the glucose concentration was increased from 15 to 20 mM; conversely, decreasing the glucose concentration from 20 to 15 mM permitted a marked increase in glucose transport activity. In this experiment the glucose concentration was then held constant at 15 mM and the citrate at 75 mM, while increasing concentrations of gluconate were introduced. The glucose transport system was progressively repressed while that for gluconate displayed a pronounced increase and was followed by that of the 2-oxogluconate uptake system.

The role of gluconate as the inhibitory agent was confirmed with a glucose dehydrogenase-negative mutant. This organism, unable to convert glucose to gluconate, displayed steadily increasing glucose uptake activity as the glucose concentration was increased in the presence of 75 mM-citrate and revealed its highest activity on 45 mM-glucose alone (Table 6). Neither the gluconate nor the 2-oxogluconate transport system was induced, as predicted from this metabolic lesion. However, the introduction of 10 mM-gluconate to the steady state on 45 mM-glucose caused substantial repression of glucose transport and induced both the gluconate and 2-oxogluconate uptake systems.

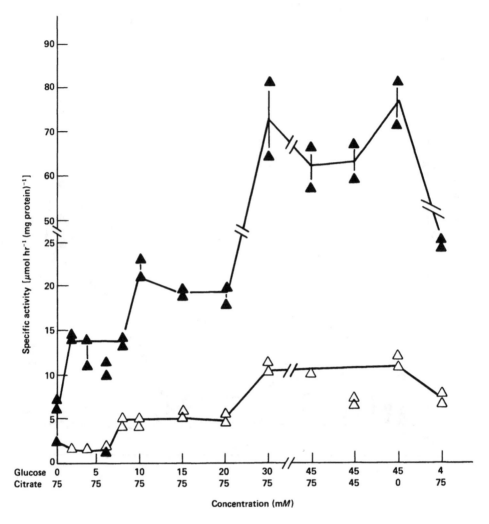

FIGURE 8.   Effect of relative concentrations of citrate and glucose in the inflowing medium on the enzymes of the extracellular pathway of glucose metabolism in ammonium-limited *Pseudomonas aeruginosa*. D = 0.40 hr⁻¹, pH 7.1. Glucose dehydrogenase, △; gluconate dehydrogenase, ▲.

E. Effect of Carbon Limitation

The effect on the three transport systems of changing the limiting nutrient from ammonium to glucose[19] is recorded in Table 7. Glucose transport was increased by some fivefold, while the gluconate and 2-oxogluconate transport systems were repressed by about sevenfold and fivefold, respectively. This behavior was also reflected by the enzymes of glucose catabolism (Table 8). Thus the enzymes associated with the extracellular pathway, namely glucose dehydrogenase, gluconate dehydrogenase, gluconate kinase, and the 2-oxogluconate enzymes (2-oxogluconate kinase and 2-oxogluconate 6-phosphate reductase) were all significantly repressed in glucose-limited cells, whereas the levels of hexokinase and glucose 6-phosphate dehydrogenase, the initial enzymes of the phosphorylative pathway, were substantially increased. The enzymes of the Entner-Doudoroff pathway (assayed together) did not display any marked change.

These results indicate that *P. aeruginosa* responds to a restriction of the glucose supply by diverting its metabolism of the available substrate very substantially from

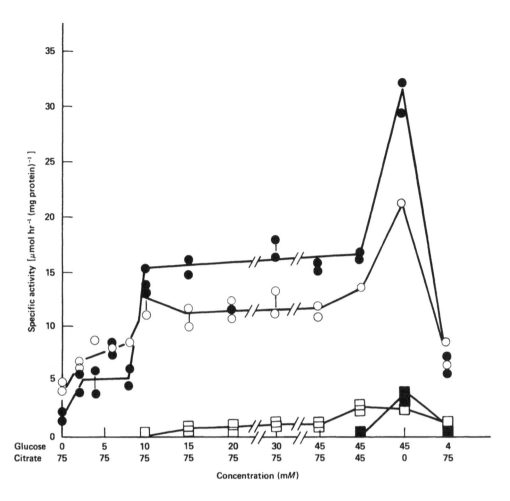

FIGURE 9.   Effect of relative concentrations of citrate and glucose in the inflowing medium on the enzymes of the intracellular pathway of glucose catabolism of ammonium-limited *Pseudomonas aeruginosa*. D = 0.40 hr^{-1}, pH 7.1. Hexokinase, O; glucose 6-phosphate dehydrogenase, ●; gluconate kinase, □; 2-oxogluconate enzymes, ■.

the extracellular direct oxidative route to the intracellular phosphorylative pathway. However, measurements of the total glucose metabolized under conditions of nitrogen and carbon limitation gave average values of 6.75 and 6.25 mmol h^{-1}(gram dry weight)$^{-1}$ respectively, indicating a decrease of only some 7% when the glucose supply was restricted. Thus the organism responds to a restricted supply of glucose by adjusting its metabolism to take the substrate into the cell as rapidly as possible, and thereby affords an excellent example of microbial response to a low-nutrient environment, as discussed by Tempest and Neijssel[20] for *Klebsiella aerogenes* in respect of both ammonium and glycerol uptake.

The extracellular oxidation of glucose to gluconate and 2-oxogluconate clearly plays an important part in the metabolism of glucose in batch culture and in NH$_4^+$-limited continuous culture. Estimates made of the flow of carbon in the extracellular pathway indicate it is tenfold that transported into the cell via the glucose transport system when the organism is presented with high glucose concentrations.[6,16]

We have suggested that the persistence of the extracellular pathway during the course of evolution has conferred the ability on *P. aeruginosa,* in its natural habitat, to se-

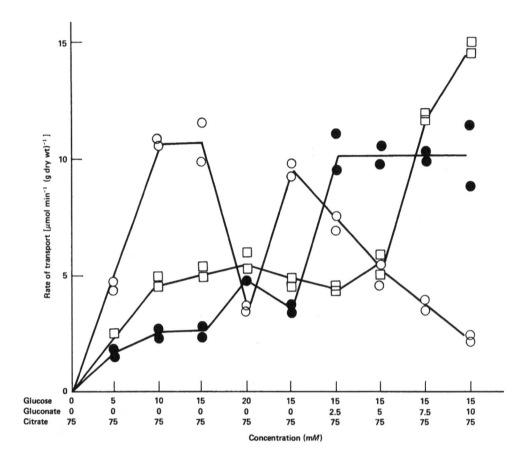

FIGURE 10.   Effect of various concentrations of citrate, glucose and gluconate in the inflowing medium on the transport systems for glucose, gluconate and 2-oxogluconate in ammonium-limited *Pseudomonas aeruginosa*. D = 0.40 hr^{-1}, pH 7.1. Transport of glucose, O; gluconate, ●; 2-oxogluconate, □.

quester glucose as gluconate and 2-oxogluconate, compounds that are not so readily utilized by various other organisms that may effectively compete for glucose.

The foregoing researches illustrate admirably the application of continuous cultivation as a very powerful weapon for the elucidation of the interplay of mixed substrates on the regulation of transport systems and enzymes of glucose metabolism. The chemostat permits the concentrations of mixed substrates to be held constant at predetermined values and their values to be altered at will. Indeed, without this technique the problems posed by such studies would have been largely insoluble.

IV. CARBON AND ENERGY RESERVE COMPOUNDS

A. General Considerations

The accumulation of carbon reserve materials such as polyglucoses (glycogen) and lipids (including poly-β-hydroxybutyrate) in microbial cells generally has been associated with the establishment of a nutrient limitation in the presence of an excess of the carbon and energy source (for a review, see Dawes and Senior[21]). Frequently, but not invariably, the nutrient limitation potentiating the deposition is the nitrogen source. Some microorganisms are able to accumulate more than one type of reserve material and, in these cases, the environmental conditions and the regulatory mechanisms in-

Table 6
THE EFFECT OF VARYING CONCENTRATIONS
OF CITRATE, GLUCOSE, AND GLUCONATE ON
THE ACTIVITIES OF THE TRANSPORT SYSTEMS
FOR GLUCOSE, GLUCONATE AND 2-
OXOGLUCONATE IN A GLUCOSE
DEHYDROGENASE-NEGATIVE MUTANT, *P.
AERUGINOSA* PAOL6

Ratio of citrate, glucose, and gluconate in inflowing medium (mM)	Rate of transport [μmol min^{-1} (g dry wt)$^{-1}$]		
	Glucose	Gluconate	2-Oxogluconate
75:10:0	2.7	n.d.	n.d.
	1.9	n.d.	n.d.
75:30:0	5.0	n.d.	n.d.
	7.6	n.d.	n.d.
75:45:0	6.0	n.d.	n.d.
	7.2	n.d.	n.d.
0:45:0	8.6	n.d.	n.d.
	9.2	n.d.	n.d.
0:45:10	4.6	9.8	24.2
	3.7	9.6	29.6

Note: Nitrogen limitation; dilution rate, 0.42 hr^{-1}; n.d., not detected. The two values are the mean of duplicate determinations performed on separate culture samples from each steady state.

From Whiting, P. H., Midgley, M., and Dawes, E. A., *Biochem. J.,* 154, 659, 1976. With permission.

Table 7
EFFECT OF LIMITING NUTRIENT ON THE
TRANSPORT ACTIVITIES OF GLUCOSE-GROWN
P. AERUGINOSA PA01

Limiting nutrient	Rate of transport [μmol min^{-1} (g dry wt)$^{-1}$]		
	Glucose	Gluconate	2-Oxogluconate
Nitrogen	2.9	19.7	47.9
(12mMNH4$^+_4$)	2.7	22.4	55.8
Carbon (30 mM glu-	15.1	2.9	9.6
cose)	14.7	3.4	11.4

Note: Dilution rate, 0.41 hr^{-1}. The two values quoted are the mean of duplicate determinations performed on two separate culture samples from each steady state.

From Whiting, P. H., Midgley, M., and Dawes, E. A., *J. Gen. Microbiol.,* 92, 304, 1976. With permission.

volved will determine the principal reserve deposited. Examples of such dual reserves include glycogen and lipid in *Mycobacterium phlei*,[22] glycogen and poly-β-hydroxybutyrate in *Rhodospirillum rubrum*[23] and *Bacillus megaterium*[24] and glycogen and polyphosphate in *Aerobacter (Klebsiella) aerogenes*[25,26] Bacillus megaterium additionally can synthesize polyphosphate under the appropriate environmental conditions.[24]

Table 8

THE EFFECT OF GLUCOSE- AND NITROGEN-LIMITED MEDIA ON THE GLUCOSE-CATABOLIZING ENZYMES OF *P. AERUGINOSA* PA01

Specific activity [μmol substrate converted hr^{-1} (mg protein)$^{-1}$]

Limiting nutrient[a]	Glucose dh	Gluconate dh	Glucose 6-phosphate dh	Hexokinase	Gluconate kinase	2-Oxogluconate metabolizing enzymes	Entner-Doudoroff enzymes	Isocitrate dh
Ammonium	10.8	69.8	29.9	22.7	2.4	2.7	19.7	119.6
	13.6	59.6	34.6	19.2	2.1	2.9	14.6	122.7
Glucose	8.5	11.1	68.0	31.0	0.7	0.3	15.1	46.0
	9.1	15.0	57.2	28.3	0.7	0.3	16.7	41.0

Note: dh, Dehydrogenase.

[a]   Nitrogen-limited medium contained 12 m*M* ammonium sulfate and 45 m*M* glucose. Glucose-limited medium contained 24 m*M* ammonium sulphate and 30 m*M* glucose. The bacterial cell density was approximately 2 mg dry wt mℓ^{-1}.

[b]   The two values quoted are the mean of duplicate determinations performed on two separate culture samples from each steady state.

From Whiting, P. H., Midgley, M., and Dawes, E. A., *J. Gen. Microbiol.*, 92, 304, 1976. With permission.

The critical importance of environmental factors in the deposition and subsequent utilization of carbon reserve compounds clearly indicates the potential value of the continuous culture technique in acquiring an understanding of the physiological, enzymological, and regulatory mechanisms that operate in the biosynthesis and degradation of these polymers. However, much of the research in this field has been carried out with batch culture; here we shall confine our attention to selected examples of the application of the chemostat approach to the problem.

B. Glycogen and Glycogen-like Polymers

The accumulation of glycogen in *Escherichia coli* was first studied in continuous culture by Holme and colleagues. [27-29] With nitrogen-limited, glucose-grown cultures of *E. coli* B, the quantity of glycogen accumulated was shown to be inversely related to the growth rate[27] (Figure 11). They observed that there were two main fractions of glycogen in this organism, one with a high molecular weight (40 to 90 × 10⁶ daltons) and the other with a low molecular weight (<2 × 10⁶ daltons) and that the major portion of the glycogen that accumulated in slow-growing bacteria was the high molecular weight fraction—while at fast growth rates, when the cell contained less glycogen, the low molecular weight fraction comprised up to half of the total glycogen content.[28,29]

C. Poly-β-hydroxybutyrate

1. Introduction

Poly-β-hydroxybutyrate, a linear homopolymer of D(−)-3-hydroxybutyric acid, is a specialized reserve of carbon and energy that accumulates in a variety of microorganisms under appropriate conditions of nutrient limitation. It is particularly widespread in the Azotobacteriaceae. While earlier studies on the deposition of this polymer were (understandably and inevitably) carried out in batch culture, it has subsequently been the focus of extensive chemostat investigations that are now considered.

2. Bacillus megaterium

The first attempt to simulate the effect of the natural physiological environment on the formation of poly-β-hydroxybutyrate appears to have been the chemostat experiments of Wilkinson and Munro[24] with *Bacillus megaterium*. They observed that *B. megaterium* KM, grown with either the nitrogen, sulfur, potassium, or carbon and energy source as the limiting factor, accumulated poly-β-hydroxybutyrate with a maximum concentration at a dilution rate of approximately 0.4 hr⁻¹ The specific growth rate had an important effect on the amount of poly-β-hydroxybutyrate accumulated as revealed by Figure 12, where it can be seen that the pattern of poly-β-hydroxybutyrate accumulation was rather similar for all the different growth-limiting conditions examined. Of particular interest is the fact that significant quantities of poly-β-hydroxybutyrat (a maximum of 12% of the dry weight) were accumulated by carbon- and energy-limited organisms, suggesting that poly-β-hydroxybutyrate biosynthesis in *B. megaterium* is not a metabolic "shunt" process whereby carbon in excess of cellular requirements is transformed to a waste product. These same experiments also reveal that significant glycogen deposition by this organism occurs only with nitrogen-limited cells and at dilution rates of 0.5 hr⁻¹ and below.

3. Alcaligenes eutrophus

Alcaligenes eutrophus H16 *(Hydrogenomonas eutropha* H16) was studied in a chemostat specially designed by Schuster and Schlegel[30] for the growth of the autotroph on limiting concentrations of hydrogen or oxygen, the gaseous components being produced by electrolysis of the mineral medium. Growth in the chemostat revealed important differences from static culture of the organism. Whereas the rate of gas uptake

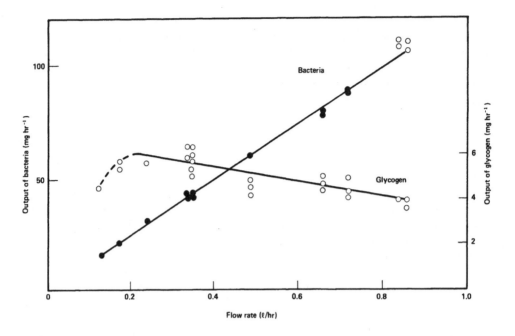

FIGURE 11.   Output of bacteria and glycogen as a function of the flow rate in a continuous culture of *Escherichia coli* B growing with ammonium limitation. Bacteria, ●; glycogen, ○. (After Holme[27]).

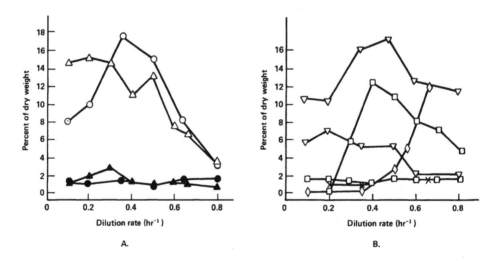

FIGURE 12.   Effect of nutrient limitation on the intracellular content of poly-β-hydroxybutyrate and glycogen of *Bacillus megaterium* KM over a range of dilution rates. Limitations were (A) potassium and sulfur, (B) carbon and energy (glucose) and nitrogen. Potassium-limited, ○; sulfur-limited, △; glucose-limited, □; glucose-limited plus excess acetate, ◊ ; nitrogen-limited, ▽; nitrogen-limited plus excess acetate, X. Open symbols represent PHB content and solid symbols glycogen content. (Adapted from Wilkinson and Munro[24]).

and hydrogenase activity declined concomitantly in static culture, in continuous culture, with hydrogen as the limiting factor, high specific activities of hydrogenase and glyceraldehyde 3-phosphate dehydrogenase were observed and the poly-β-hydroxybutyrate content was negligible. However, when oxygen became the limiting factor for growth the organisms accumulated up to 23% of their dry weight of poly-β-hydroxybutyrate and the specific activites of hydrogenase and glyceraldehyde 3-phosphate dehydrogenase were low. Thus *Alcaligenes eutrophus* was shown to accumulate poly-β-hydroxybutyrate under conditions of oxygen limitation, as well as under the previously observed nitrogen limitation in batch culture.[31]

4. Azotobacter beijerinckii
a. Batch Culture Experiments

A prelude to our own researches on poly-β-hydroxybutyrate metabolism was a survey[32] of a large number of Azotobacter species and the choice for subsequent detailed biochemical investigation of a particular strain of *Azotobacter beijerinckii* that was capable of accumulating up to 75% of its dry weight of the polymer. Batch cultures of this obligate aerobe growing in glucose media and fixing nitrogen accumulated poly-β-hydroxybutyrate towards the end of exponential growth and during the stationary phase.[33] With 0.5% (w/v) glucose the maximum yield of polymer was 35% (w/w) and, following the exhaustion of the carbon source, synthesis of polyβ-hydroxybutyrate ceased and it was rapidly degraded without a concomitant decrease in total cell dry weight. When 2% (w/v) glucose was included in the medium, polymer was again deposited towards the end of the exponential growth, but, unlike the pattern with limiting glucose, polymer accumulation continued during the stationary phase and at the final analysis represented 74% of the dry weight.

Experiments with a Mackereth oxygen electrode inserted in batch cultures revealed that poly-β-hydroxybutyrate biosynthesis was initiated at about the time when the dissolved oxygen concentration was zero, which coincided with the transition from exponential growth. These observations led us to consider that oxygen limitation might be the critical factor that initiated poly-β-hydroxybutyrate synthesis in *A. beijerinckii*. However, in a batch culture simultaneously fixing atmospheric nitrogen and utilizing atmospheric oxygen, an apparent oxygen limitation could well be masking a real nitrogen limitation and it was essential therefore to separate these two parameters. This was achieved by recourse to chemostat experiments.

b. Chemostat for Studies of Nitrogen-Fixing A. beijerinckii

The chemostat used was a Porton-type apparatus[34] of 2-liter capacity, which possessed sufficient ports to accommodate probes for measurement and control of pH, temperature (heating was effected by an external 250 W infrared lamp and the vessel was fitted with an internal water coil), redox potential, oxygen concentration, medium inflow, gas inflow and effluent, culture effluent, and ports for inoculation, sampling of culture, antifoam addition, and vessel drainage. The culture vessel, minus oxygen electrode (but otherwise complete) was autoclaved at 20 lb in.$^{-2}$ (138 kPa) for 1 hr. The oxygen electrode was sterilized by immersion for 1 hr in 10% (v/v) formaldehyde solution before insertion into the culture vessel. Medium was pumped into the vessel by a H.R. flow-inducer (Watson-Marlow Ltd., Marlow, Bucks, U.K.). To prevent back-growth from the culture vessel into the medium supply-lines a hot-water jacket (60°C) surrounded the medium line immediately before the inlet port.

All gases were obtained in cylinders and gas flows were measured by rotameters. (Series 1100, 300 mm type; M. F. G. Rotameter Co. Ltd., Croydon, Surrey, U.K.) and regulated by Flo-stat regulators (G. A. Platon Ltd., Basingstoke, Hants., U.K.) at a working pressure of 10 lb in.$^{-2}$ (69 kPa). The combined gases were passed through two Mackley filters (Microflow Ltd., Mackley Filter Division, Gateshead 8, Co. Durham, U.K.; NaCl penetration less than 0.001%), arranged in series, before flowing into the culture vessel. Effluent gases were passed through a cold-water condenser and a packed cottonwool filter before measurement of total effluent gas flow by a rotary gas-meter (Lange, Gelsenkirchen; Fisons Scientific Apparatus Ltd., Loughborough, Leics, U.K.). The effluent gas was dried by passage through a silica gel column (2.5 cm × 30 cm) and was then divided for analysis of oxygen by a paramagnetic analyser (Servomex Oxygen Analyser, Type OA 137; Servomex Controls Ltd., Crowborough, Sussex, U.K.) and carbon dioxide by an M.S.A. infrared analyser (Lira Model 300; Mine Safety Appliances Co. Ltd., Glasgow E.3, U.K.).

A Mackereth[35] oxygen electrode was used to determine the dissolved oxygen concentration in the culture. Before inoculation the culture medium was saturated with air and the maximum current output from the electrode recorded as 100% of air-saturation; oxygen-free nitrogen was passed through the medium to set the oxygen concentration at zero. The oxygen concentration of the culture could be controlled to any value between 0 and 100% air-saturation by means of a series 60 controller (Leeds and Northrup Ltd., Birmingham, U.K.) by using a 3-action C.A.T. control unit. The control was based on the method first described by MacLennan and Pirt.[36] Several modifications were necessary because of the nitrogen-fixing capacity of our organism. The electro-pneumatic transducer, operated by compressed air (20 lb in.$^{-2}$; 138 kPa), fed its output signal to a control valve through which pure oxygen was passed. If compressed air was used as the source of oxygen, attainment of a nitrogen-limited culture was impossible because of the irregularity of nitrogen gas flows.

Culture redox potential (E_h) was measured with a smooth platinum spade-electrode (Activion Ltd., Kinglassie, Fife, U.K.) by using a KCl salt bridge and a calomel reference electrode. The steam-sterilizable electrode was calibrated by the method of Jacob.[37]

Output current was measured by a high-resistance pH-meter (model 91B; Electronic Instruments Ltd., Richmond, Surrey, U.K.) to avoid polarity reversal at the electrode.

c. *Effect of Various Nutrient Limitations and Growth Rate on Poly-β-hydroxybutyrate Accumulation*

The organism was grown with either carbon, oxygen, or nitrogen limitation over a range of dilution rates (Figure 13). Clearly a massive deposition of the polymer occurred only under conditions of oxygen limitation. Cultures grown with nitrogen or carbon-limitation rarely contained more than 3% of their dry weight as poly-β-hydroxybutyrate, whereas oxygen-limited cultures displayed values ranging from 20% at the highest specific growth rate (D = 0.252 hr^{-1}) to 45% at the lowest rate (D = 0.049 hr^{-1}). The very high contents of polymer recorded in batch cultures (65 to 74%) were not attained under the conditions of oxygen limitation used in these experiments. However, this difference was subsequently shown to be a function of the degree of oxygen limitation imposed in the chemostat,[38] and that these very high polymer contents could be observed if the oxygen supply rate was taken well below that at which the oxygen electrode registered zero dissolved oxygen concentration. Presumably only at very low oxygen supply rates did the chemostat cultures simulate the conditions of oxygen limitation that obtain towards the end of exponential growth in batch culture, and this applied with both nitrogen-fixing and ammonium-grown cultures.

The variation of yield (Y_1) of bacteria [g(mol of glucose utilized)$^{-1}$] and of poly-β-hydroxybutyrate $[Y_2;$ g(mol of glucose utilized)$^{-1}$] with growth rate is recorded in Tables 9 and 10. The difference $(Y_1 - Y_2)$ represents a yield value relating cell mass, other than poly-β- hydroxybutyrate, to glucose consumed. In Tables 9 and 10 the value of $Y_{av\ e^-}$ represents the yield of cells as gram per available electron *(av e⁻)*; glucose was assumed to have an *av e⁻* value of 24.

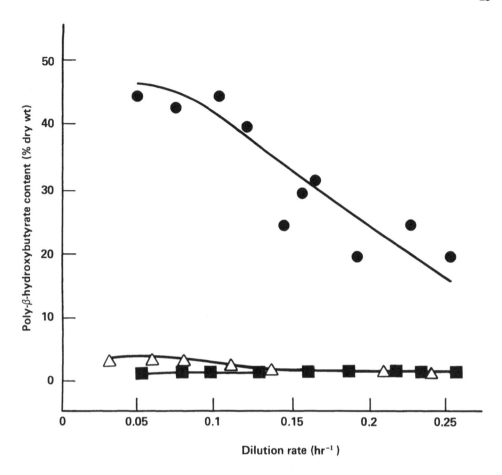

FIGURE 13.   Effect of different nutrient limitations and dilution rate on the poly-β-hydroxybutyrate content of *A. beijerinckii*. Limitation: glucose, △; nitrogen, ■; oxygen, ●.

Table 9

VARIATION OF BIOMASS YIELD AND POLY-β-HYDROXYBUTYRATE CONTENT OF *A. BEIJERINCKII* WITH GROWTH RATE IN NITROGEN-LIMITED CHEMOSTAT CULTURES

Growth rate D (hr^{-1})	Yield of biomass [g (mol of glucose)$^{-1}$] (Y$_s$)	Y av e$^-$	Poly-β-hydroxybutyrate content [% (w/w) of biomass]	Bacterial doubling time, t_d(hr)
0.053	7.4	0.31	<1.5	13.07
0.079	10.8	0.45	<1.5	8.77
0.098	10.7	0.44	<1.5	7.11
0.130	12.5	0.52	<1.5	5.29
0.161	14.4	0.60	<1.5	4.31
0.186	16.3	0.68	<1.5	3.73
0.218	18.7	0.78	<1.5	3.18
0.234	19.7	0.82	<1.5	2.96
0.255	Culture washed out			2.72

Note: Abbreviation; Y av e$^-$,g of bacteria/available electron.

From Senior, P. J., Beech, G. A., Ritchie, G. A. F., and Dawes, E. A., *Biochem. J.*, 128, 1193, 1972. With permission.

Table 10

VARIATION OF BIOMASS YIELD AND POLY-β-HYDROXYBUTYRATE CONTENT
OF *A. BEIJERINCKII* WITH GROWTH RATE IN OXYGEN-LIMITED CHEMOSTAT
CULTURES

Growth rate $D(hr^{-1})$	Yield of biomass [g(mol of glucose)$^{-1}$] (Y_1)	Y av e$^-$	Yield of poly-β-hydroxybutyrate [g (mol of glucose)$^{-1}$] (Y_2)	Y_1-Y_2	Poly-β-hydroxybutyrate content [% (w/w) of biomass]	Bacterial doubling time t_c (hr)
0.049	50.9	2.12	22.6	28.3	45	14.14
0.075	49.7	2.07	21.4	28.3	43	9.24
0.102	48.4	2.02	21.6	26.8	45	6.79
0.121	55.7	2.32	22.1	33.6	40	5.71
0.146	70.1	2.92	17.6	52.5	25	4.75
0.158	64.0	2.66	19.3	44.7	30	4.40
0.164	49.0	2.04	15.6	33.4	32	4.22
0.192	52.9	2.20	10.6	42.3	20	3.61
0.277	63.3	2.64	16.1	47.2	25	3.05
0.252	55.3	2.31	10.7	44.6	20	2.75

Note: See also Table 9.

From Senior, P. J., Beech, G. A., Ritchie, G. A. F., and Dawes, E. A. *Biochem. J.*, 128, 1193, 1972. With permission.

d. Effects of Imposition and Relaxation of Oxygen Limitation

The effect of the sudden imposition of an oxygen limitation on a nitrogen-limited chemostat culture is shown in Figure 14. The nitrogen-limited culture ($D = 0.233$ h^{-1}) was grown with the dissolved oxygen concentration controlled at 10% of air saturation. On decreasing the oxygen supply rate to 0.25 and then 0.15 of its original value, the dissolved oxygen concentration decreased rapidly to zero and was accompanied by a decrease in E_h from +15 mV to −50 mV. At this point polymer accumulation commenced, and the E_h value, after an initial oscillation, rose to +30 mV. Meanwhile the dissolved oxygen concentration remained at zero and the glucose concentration in the culture increased from 26 to 70 mM. The maximum value for polymer content, approximately 45% (w/w) of biomass, was not maintained in this oxygen-limited culture and fell to approximately 20% of the biomass after 32 hr, a value characteristic of the steady state poly-β-hydroxybutyrate content at this growth rate (cf. Table 10). The high poly-β-hydroxybutyrate content initially observed in response to the sudden imposition of an oxygen limitation thus appears to be a transient phenomenon, at fast growth rates.

The converse experiment of relaxing an oxygen limitation with the subsequent transition to a carbon-limited culture was also carried out (Figure 15).[39] Before relaxation of oxygen limitation the culture was controlled at 30°C and pH 6.6 with the dissolved oxygen concentration at less than 0.5% of air saturation. The redox potential E_h was +50 mV, the steady-state dry weight of culture 1.95 mg mℓ^{-1}, culture supernatant glucose concentration 37.6 mM, poly-β-hydroxybutyrate content 48% of the dry biomass and the dilution rate 0.10 hr^{-1}. At zer time the oxygen concentration was increased to 10% of air saturation, attainment of which took 2 min; thereafter this concentration was maintained for the rest of the experiment. E_h became more positive and the culture dry weight and poly-β-hydroxybutyrate content decreased (Figure 15). The glucose concentration, after an initial decrease during the first 30 min after relaxation of the oxygen limitation (Figure 16) remained constant for a further 1.5 hr. For the next 7 hr, E_h became more negative, and the culture dry weight decreased together with the glucose concentration (Figures 15 and 16).

At 11.5 hr the culture became glucose-limited and the poly-β-hydroxybutyrate content was then 6% (w/w) of the dry weight. At the point of glucose limitation (arrowed in Figure 15), E_h became more positive (increasing from +50 mV to +120 mV) and remained between +110 mV and +120 mV for the remainder of the experiment. During the 4 hr period before glucose limitation, the culture went through a period of batch growth and, by the onset of glucose limitation, had overshot the value of the steady-state dry weight dictated by the continuing supply of glucose. Thus, during the period 15 to 29 hr, the culture dry weight fell from 1.6 to 1.03 mg mℓ^{-1}, the latter value representing the dry weight of a new, glucose- limited steady state (with inflowing medium glucose concentration of 55 mM). After 13 hr the intracellular poly-β-hydroxybutyrate content was <1.5% of the dry weight and remained at this low value for the rest of the experiment

These chemostat experiments thus indicated that poly-β-hydroxybutyrate accumulation in *A. beijerinckii* occurred in response to an oxygen limitation. Although the pitfalls of interpretation of the observations made with a redox electrode inserted into such a heterogeneous system as a bacterial culture are patent, we suggested that they were, nonetheless, in accord with our concept of poly-β-hydroxybutyrate accumulation serving as an electron sink for excess reducing power that accumulated when the cell became oxygen limited and electron transport to oxygen via the electron transport chain was restricted. Evidence to support this hypothesis was provided by the discovery that oxygen-limited cultures, which displayed very low rates of oxygen utilization and

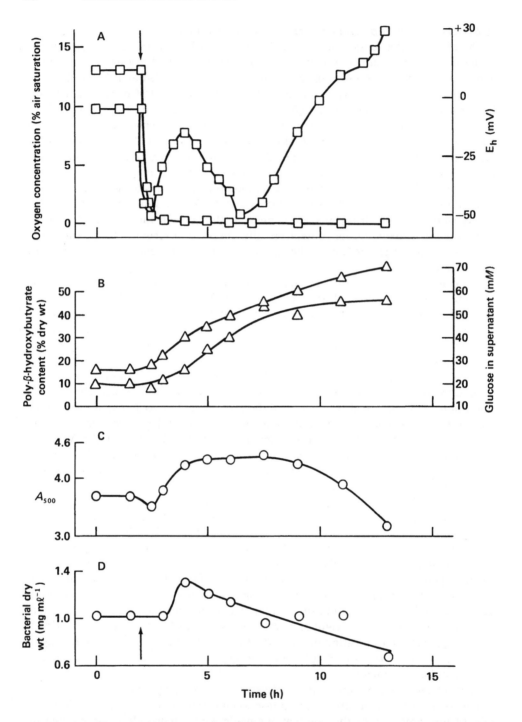

FIGURE 14.   Effect of imposition of oxygen limitation on a nitrogen-limited chemostat culture of *A. beijerinckii*. The nitrogen-limited dilution rate was 0.233 hr^{-1} and, at the point indicated by the arrow, the oxygen supply rate was decreased from 100 mℓ min^{-1} to 25 mℓ min^{-1}, imposing an oxygen limitation; 40 min later the oxygen supply rate was decreased further to 15 mℓ min^{-1}. (A) Culture dissolved oxygen concentration, □; redox potential *in situ*, ■; (B) poly-β-hydroxybutyrate content, △; glucose concentration in culture, ▲; (C) culture absorbance (A$_{500}$), ●; (D) bacterial dry weight m$\ell$$^{-1}$, ○.

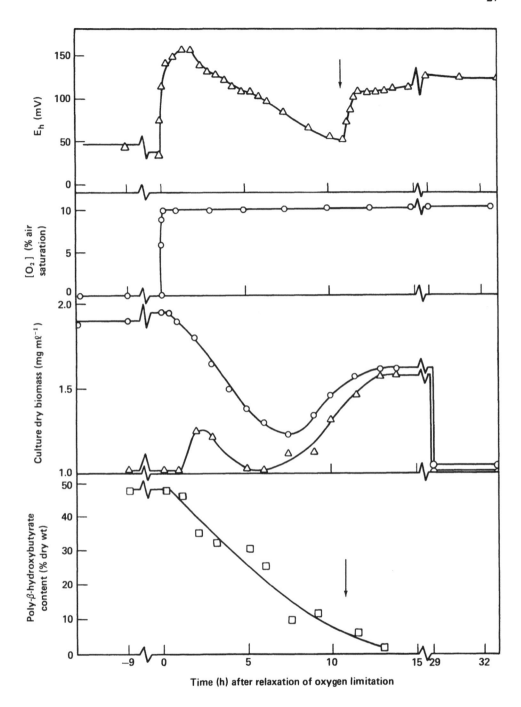

FIGURE 15. Effect of relaxation of oxygen limitation on an oxygen-limited chemostate culture of *A. beijerinckii*. The oxygen-limited culture dilution rate was 0.10 hr⁻¹ and, at zero time, the oxygen supply rate was increased automatically by the dissolved oxygen controller to maintain a dissolved oxygen concentration of 10% of air saturation. The arrows indicate the time when the glucose in the culture became exhausted. (A) Redox potential *in situ*, $\triangle$; (B) culture dissolved oxygen concentration, $\bullet$; (C) culture dry weight, $\bigcirc$; culture dry weight minus poly-β-hydroxybutyrate content, $\blacktriangle$; (D) bacterial poly-β-hydroxybutyrate content, $\square$.

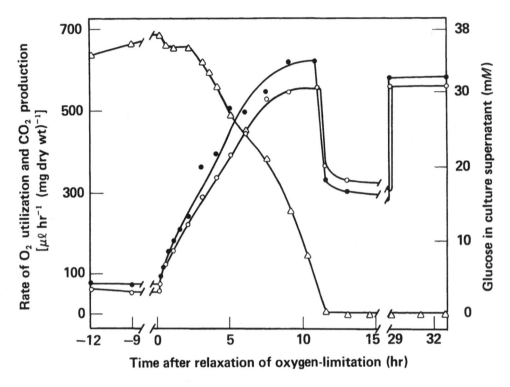

FIGURE 16.   Effect of relaxation of oxygen limitation on an oxygen-limited chemostat culture of *A. beijerinckii.* Experimental details as for Figure 15. Culture glucose concentration, △; rate of oxygen utilization *in situ,* ● rate of evolution of carbon dioxide *in situ,* ○.

carbon dioxide formation *in situ* compared with the corresponding rates for nitrogen and carbon-limited cultures, had lower potential rates of gas exchange, i.e., when oxidation was tested with glucose in the presence of excess oxygen, oxygen-limited cells consumed oxygen at barely one-seventh the rate of glucose-limited organisms.

Figure 17 shows the change in $NADH/NAD^+$ ratio associated with the imposition of an oxygen limitation on a nitrogen-limited chemostat culture operated at a dilution rate of $0.1 \, hr^{-1}$. The NADH concentration increased rapidly for the first 30 min; after about 90 min the NADH/NAD ratio was restored to a slightly higher value than that of the previous nitrogen-limited steady state.[40] By this time poly-β-hyroxybutyrate synthesis was proceeding rapidly and thus serving as an electron acceptor alternative to oxygen. The total NAD(H) concentrations remained constant at 1.92 nmol NAD(H) (mg bacterial protein)$^{-1}$.

e. Regulation of Enzyme Activities by Oxygen Concentration

The critical role of oxygen concentration in the initiation of polymer synthesis was investigated at the enzymic level initially by studying the effect of NAD(P)H on various enzymes of carbon metabolism. In *A. beijerinckii* glucose is metabolized principally via the Entner-Doudoroff pathway and the tricarboxylic acid cycle. Senior and Dawes[41] found that glucose 6-phosphate dehydrogenase, citrate synthase, and isocitrate dehydrogenase were all powerfully inhibited by either or both of NADH and NADPH, so that under conditions of oxygen limitation, when the concentration of these reduced coenzymes increased, glucose metabolism and the operation of the tricarboxylic acid cycle, which generates intermediates and energy for biosynthesis, would be decreased.

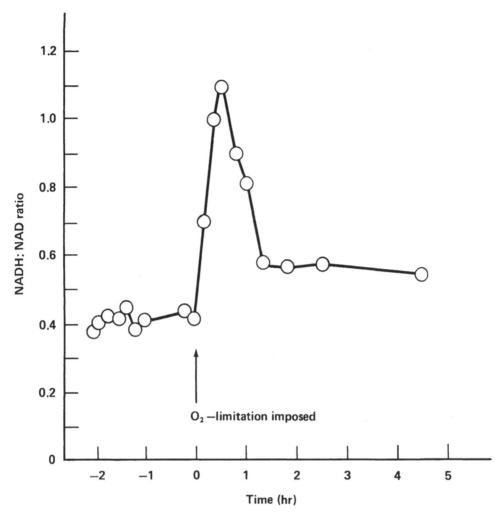

FIGURE 17. Effect on the NADH/NAD⁺ ratios of *A. beijerinckii* of imposition of an oxygen limitation on a nitrogen-limited culture. Oxygen limitation (zero d.o.t.) was imposed at zero time. $D = 0.10 \, hr^{-1}$.

The serious consequences of such inhibitory effects for growth can be largely offset by some acetyl-CoA being diverted from entry to the tricarboxylic acid cycle to poly-β-hydroxybutyrate synthesis, the reductive step of which partially alleviates the accumulated reducing power and so permits glucose metabolism and the tricarboxylic acid cycle to continue to function at a rate greater than would otherwise be possible.

f. Poly-β-hydroxybutyrate Metabolism and Its Regulation

The route of poly-β-hydroxybutyrate synthesis involves the condensation of two molecules of acetyl-CoA, catalysed by β-ketothiolase, to release CoA and form aceto-acetyl-CoA, which is then reduced to D(−)-3-hydroxybutyryl-CoA by 3-hydroxybutyryl-CoA dehydrogenase, an enzyme which utilized NADPH at some fivefold the rate of NADH.[42] 3-Hydroxybutyryl-CoA is then the substrate for a granule-membrane-bound polymerase, which simultaneously releases CoA.

$$2\text{-Acetyl-CoA} \xrightleftharpoons{\beta\text{-ketothiolase}} \text{acetoacetyl-CoA} + \text{CoA}$$

$$\text{Acetoacetyl-CoA} + \text{NADPH} + \text{H}^+ \xrightleftharpoons[]{\substack{\text{3-hydroxybutyrate}\\ \text{dehydrogenase}}} \text{D}(-)\text{-3-hydroxybutyryl-CoA} + \text{NADP}^+$$

$$(\text{PHB})_n + \text{D}(-)\text{-3-hydroxybutyryl-CoA} \xrightarrow{\text{polymerase}} (\text{PHB})_{n+1} + \text{CoA}$$

Degradation of the polymer occurs via a different route that does not involve CoA. A granule-bound depolymerase releases $D(-)$-3-hydroxybutyrate, which is oxidized via an NAD-specific dehydrogenase to acetoacetate; succinyl-CoA transferase then converts it to acetoacetyl-CoA. The discovery of this latter enzyme revealed that poly-β-hydroxybutyrate metabolism was a cyclic process with acetyl-CoA functioning both as a precursor and a product, and with acetoacetyl-CoA serving as an intermediate common to biosynthesis and degradation (Figure 18). Regulation is thus of paramount importance to prevent futile cycling, and Senior and Dawes[39] found that the key enzymes were β-ketothiolase and 3-hydroxybutyrate-CoA dehydrogenase. Thus β-ketothiolase, which has a high K_m of 0.9 mM for acetyl-CoA, was inhibited in the condensation reaction by free CoA and in the thiolysis reaction by acetoacetyl-CoA, although the latter inhibition could be overcome by increasing the concentration of CoA.

During conditions of unrestricted growth in the presence of excess oxygen (i.e., no polymer accumulation), the steady- state concentration of CoA would be expected to be high, mediated by the action of citrate synthase, with citrate formation serving as a sink for acetyl groups and simultaneously releasing free CoA. The combination of high CoA and low acetyl-CoA concentration would prevent acetoacetyl-CoA formation and thus poly-β-hydroxybutyrate would not be formed. The relief by CoA of acetoacetyl-CoA inhibition of acetoacetyl-CoA thiolysis would, in turn, ensure that only when CoA was present at high concentrations would degradation proceed.

When an oxygen limitation is imposed, restriction of citrate synthase activity as a result of NADH accumulation would occur, the concentration of acetyl-CoA would increase with a concomitant decrease in the concentration of CoA. These conditions lead to saturation of β-ketothiolase and its release from the inhibitory effect of CoA so that acetoacetyl-CoA synthesis will occur and poly-β-hydroxybutyrate synthesis proceed, the reductive step utilizing some of the accumulated reducing power.

There is, as yet, no detailed information available concerning the regulation of the poly-β-hydroxybutyrate depolymerizing enzyme(s) associated with the membrane of the polymer granules. Consequently there is some doubt, but it seems possible that regulation of polymer degradation could be achieved by the inhibition of $D(-)$-3-hydroxybutyrate dehydrogenase by NADH (competing with NAD$^+$), pyruvate, and 2-oxoglutarate (competing with 3-hydroxybutyrate) for this enzyme is competitively inhibited by these compounds. The regulation ofthe *A. bieijerinckii* enzyme by pyruvate and 2-oxoglutarate is probably exerted when glucose catabolism and the tricarboxylic acid cycle are operating, and the need to oxidize accumulated polyer, in its role as a reserve of carbon and energy, is minimal. A finer control is exerted by the competitive inhibition by NADH, rendering the enzyme very sensitive to changes in the NADH/NAD$^+$ ratio brought about, for example, by the imposition or relaxation of oxygen limitation of growth.

This scheme of fine control of poly-β-hydroxybutyrate metabolism thus accords with the recognized features for the deposition and utilization of a reserve material, ensuring that accumulation only occurs when the supply of exogenous carbon is in excess of the requirements for growth and maintenance, and that degradation only takes place

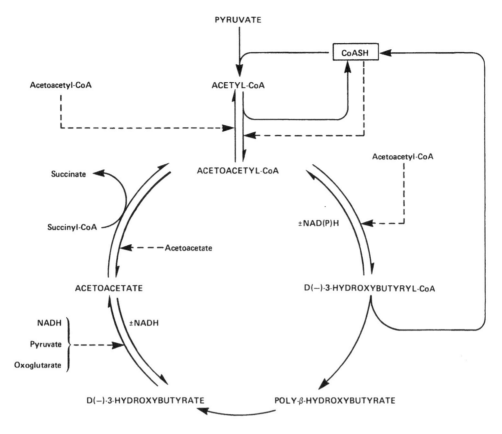

FIGURE 18. Regulation of poly-β-hydroxybutyrate metabolism in *A. beijerinckii*. Inhibition is indicated by broken lines.

when the supply of exogenous carbon is limited or exhausted. However, it was important to know also to what extent coarse control occurs, reflected by changes in the levels of enzymes of carbon metabolism, and to this end further chemostat experiments were carried out.

g. Regulation of the Tricarboxylic Acid Cycle

The effect of oxygen concentration on the levels of enzymes of poly-β-hydroxybutyrate metabolism and the tricarboxylic acid cycle were studied in transitions from oxygen to nitrogen-limitation and *vice-versa*.[40] Since (even when the organisms are nitrogen-limited) oxygen concentration is of importance in relation to the respiratory protection of the nitrogenase system,[43,44] two dissolved oxygen concentrations were chosen for nitrogen-limited cultures, one designed to supply a restricted excess of oxygen and the other a higher concentration.

Figure 19A shows the transition from oxygen limitation (zero dissolved oxygen tension) to nitrogen limitation with 0.8 kPa (6 mm) d.o.t., i.e., a restricted excess of oxygen. The poly-β-hydroxybutyrate content fell from 50 to 10% of the dry weight and β-ketothiolase and acetoacetyl-CoA reductase showed an initial decrease and then a recovery to new, lower steady-state levels. Then the d.o.t. was doubled to 1.6 kPa and (Figure 19B) the polymer content fell to a negligible value, and β-ketothiolase and acetoacetyl-CoA reductase fell further, although both were still present under these highly aerobic conditions when poly-β-hydroxybutyrate synthesis was minimal. Finally (Figure 19C), an oxygen limitation was reimposed upon the culture when the activities

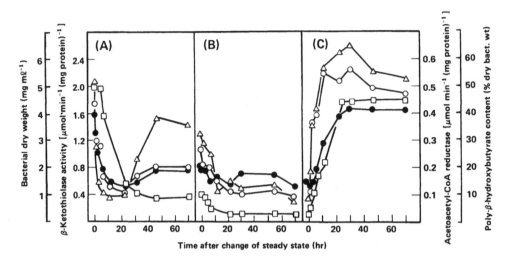

FIGURE 19.   Effect of transition from oxygen to nitrogen limitation, and vice versa, on a chemostat culture of *A. beijerinckii* on β-ketothiolase (△) and acetoacetyl-CoA reductase (○) levels, poly-β-hydroxybutyrate content (□) and bacterial dry weight (●). D = 0.10 hr⁻¹. At zero time the cultural conditions were changed from (A) oxygen limitation (zero d.o.t.) to nitrogen limitation (d.o.t. 0.8 kPa); (B) nitrogen limitation (d.o.t. 0.8 kPa) to nitrogen limitation (d.o.t. 1.6 kPa); (C) nitrogen limitation (d.o.t. 1.6 kPa) to oxygen limitation.

of both enzymes increased rapidly and was mirrored by a marked increase in poly-β-hydroxybutyrate content. Changes in the dry weight of the culture essentially reflected the changes in intracellular poly-β-hydroxybutyrate content.

The changes in NADH oxidase, isocitrate dehydrogenase, and 2-oxoglutarate dehydrogenase for the same transitions are recorded in Figure 20. The level of NADH oxidase rose rapidly for the first 12 hr after relaxation of the oxygen limitation, but then declined to a steady-state value of about 2.5 times that of the oxygen-limited organisms. Increasing the d.o.t. from 0.8 to 1.6 kPa caused a further increase in the steady-state NADH oxidase activity. *Azotobacter* species are noted for their extremely high levels of NADH oxidase, which is associated with the process of oxygen scavenging in the respiratory protection of nitrogenase, and the level of this enzyme reflects the environmental oxygen concentration. Thus the re-imposition of an oxygen limitation (Figure 20C) resulted in an immediate, marked decrease in the NADH oxidase level. Isocitrate dehydrogenase and 2-oxoglutarate dehydrogenase both increased significantly on relaxation of the oxygen limitation (Figure 20A), and the latter enzyme showed a further substantial increase when the d.o.t. was raised from 0.8 to 1.6 kPa (Figure 20B). The levels of both enzymes were restored to their original values when an oxygen limitation was re-imposed (Figure 20C). In contrast, the levels of citrate synthase and pyruvate dehydrogenase were relatively unaffected by the transitions (Figure 21).

Q_{O_2} values, glucose yield coefficients *(Y_G;* gram organism formed per mol glucose utilized) and supernatant glucose concentrations were also measured (Figure 22) for these transitions. There was an inverse relationship between Q_{O_2} and *Y_G*. After relaxation of the oxygen limitation, the supernatant glucose concentrations rose to a maximum at 22 hr before declining to a lower value as the new steady state was approached (Figure 22A). Increasing the oxygen concentration (Figure 22B) caused a decrease in glucose concentration, indicating that glucose oxidation increased as a respiratory protection function. Conversely, re-imposition of an oxygen limitation led to a rise in supernatant glucose concentration from about 20 to 82 m*M*.

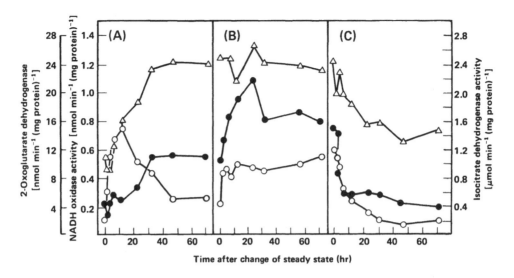

FIGURE 20.   Effect of transition from oxygen to nitrogen limitation, and vice versa, on a chemostat culture of *A. beijerinckii* on NADH oxidase (O), 2-oxoglutarate dehydrogenase (●) and isocitrate dehydrogenase (△) levels. Experimental details as for Figure 19.

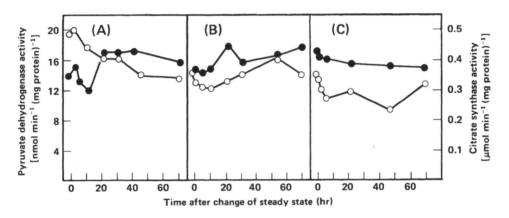

FIGURE 21.   Effect of transition from oxygen to nitrogen limitation, and vice versa, on a chemostat culture of *A. beijerinckii* on citrate synthase (●) and pyruvate dehydrogenase (O) levels. Experimental details as for Figure 19.

The regulation of tricarboxylic acid cycle activity by oxygen concentration is well documented for facultative organisms such as *Escherichia coli* (Amarasingham and Davis[45]), but it remains to be ascertained whether the changes observed with *A. beijerinckii* are generally representative of obligate aerobes or whether they are a manifestation of respiratory protection and therefore characteristic only of nitrogen-fixing organisms.

h. Regulation of Some Enzymes of Glucose Metabolism

The steady-state levels of some enzymes of glucose and poly-β-hydroxybutyrate metabolism, over a wide range of oxygen supply rates that span oxygen limitation and oxygen excess, are shown in Figure 23. In the region of oxygen limitation, increasing the oxygen supply caused the levels of β-ketothiolase and acetoacetyl-CoA reductase to decrease in parallel with the poly-β-hydroxybutyrate content of the organisms, con-

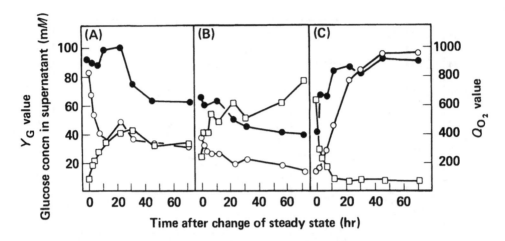

FIGURE 22.. Effect of transition from oxygen to nitrogen limitation, and vice versa, on a chemostat culture of *A. beijerinckii* on supernatant glucose concentration (●), Q_{o_2} (□) and observed Y_G (○) values. Experimental details as for Figure 19.

firming the behavior established in the transition experiments of Figure 19. The Entner-Doudoroff enzymes (gluconate 6-phosphate dehydratase and 2-oxo-3-deoxygluconate 6-phosphate aldolase assayed together) also decreased while glucose 6-phosphate dehydrogenase remained constant. The corresponding Q_{o_2}, Q_{co_2} and R.Q. values for the same series of steady states are shown in Figure 24, although it should be noted that the R.Q. values at the lowest oxygen supply rates are subject to considerable experimental error.[49] These results contrast with those of Haaker and Veeger[46] for *Azotobacter vinelandii* for they found that, with single steady states, glucose b-phosphate dehydrogenase activity was about 3.5-fold higher in nitrogen-limited than in oxygen-limited cells and the Entner-Doudoroff enzymes were slightly increased.

Nagai and colleagues[47,48] have studied the response of four enzymes of glucose metabolism in *Azotobacter chroococcum* to wide ranges of dissolved oxygen concentration, and found that only fructose 1,6-bisphosphate aldolase increased with increasing oxygen concentration, whereas isocitrate dehydrogenase, glyceraldehyde 3-phosphate dehydrogenase, and isocitrate lyase remained essentially constant. However, the range of oxygen concentrations where isocitrate dehydrogenase displayed marked changes in *A. beijerinckii* (0 to 1.0 kPa; Figure 20) was not adequately covered in their experiments, and so it is possible that they missed the region where a change might have occurred.

V. CONCLUSION

The foregoing examples of the application of the chemostat to the investigation of widely different aspects of bacterial carbon metabolism will, it is hoped, emphasize the value and versatility of the technique for researches on bacterial biochemistry and physiology.

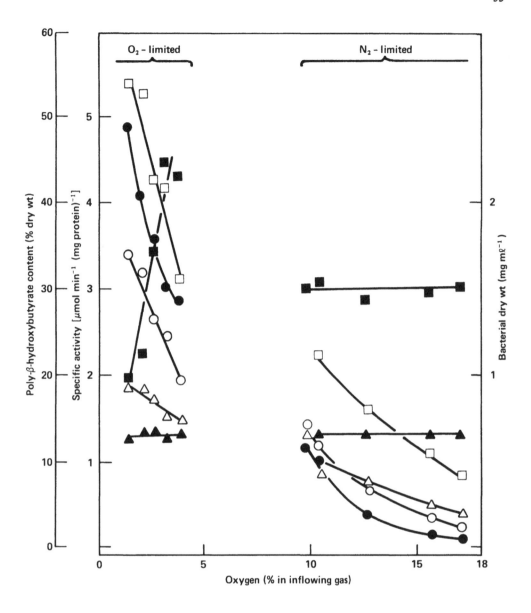

FIGURE 23.   Effect of oxygen supply in a chemostat culture of *A. beijerinckii* on the steady state levels of β-ketothiolase (□), acetoacetyl-CoA reductase, (O); glucose 6-phosphate dehydrogenase (△) and Entner-Doudoroff enzymes (▲), and on poly-β-hydroxbutyrate content (●) and bacterial dry weight (□). D = 0.10 hr⁻¹. For oxygen-limited steady states the nitrogen flow rate was 800 mℓ min⁻¹; for nitrogen-limited conditions the nitrogen flow rate was 20 mℓ min⁻¹, and the oxygen-nitrogen mixture was diluted with 591 mℓ argon min⁻¹.

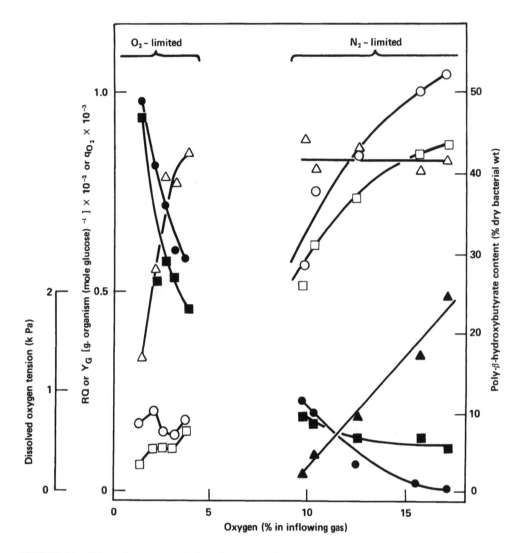

FIGURE 24.   Effect of oxygen supply in a chemostat culture of *A. beijerinckii* on growth yield on glucose (Y_G, ■), metabolic quotients q_{O_2} (O) and q_{CO_2} (□), respiratory quotient R.Q. (△), dissolved oxygen tension (▲) and poly-β-hydroxybutyrate content (●). Experimental details were as for Figure 23.

REFERENCES

1. **Harder, W. and Dijkhuisen, L.** *Continuous Culture 6: Aplications and New Fields*, Ellis Horwood Ltd., Chichester, 1976, chap. 23.
2. **Stokes, F. N. and Campbell, J. J. R.**, The oxidation of glucose and gluconic acid by dried cells of *Pseudomonas aeruginosa, Arch. Biochem. Biophys.*, 30, 121, 1951.
3. **Claridge, C. A. and Werkman, C. H.**, Formation of 2-ketogluconate from glucose by a cell-free preparation of *Pseudomonas aeruginosa, Arch. Biochem. Biophys.*, 47, 99, 1953.
4. **Claridge, C. A. and Werkman, C. H.**, Evidence for alternate pathways for the oxidation of glucose by *Pseudomonas aeruginosa, J. Bacteriol.*, 68, 77, 1954.
5. **Wang, C. H., Stern, I. J., and Gilmour, C. M.**, The catabolism of glucose and gluconate in *Pseudomonas* species, *Arch. Biochem. Biophys.*, 81, 489, 1959.
6. **Roberts, B. K., Midgley, M., and Dawes, E. A.**, The metabolism of 2-oxogluconate by *Pseudomonas aeruginosa, J. Gen. Microbiol.*, 78, 319, 1973.
7. **Blevins, W. T., Feary, T. W., and Phibbs, P. V., Jr.**, 6-Phosphogluconate dehydratase deficiency in pleiotropic carbohydrate-negative mutant strains of *Pseudomonas aeruginosa, J. Bacteriol.*, 121, 942, 1975.
8. **Tiwari, N. P. and Campbell, J. J. R.**, Enzymatic control of the metabolic activity of *Pseudomonas aeruginosa* grown in glucose or succinate media, *Biochim. Biophys. Acta*, 192, 395, 1969.
9. **Sissons, A., Whiting, P. H., Midgley, M., and Dawes, E. A.**, The pathway of gluconeogenesis in *Pseudomonas aeruginosa, Proc. Soc. Gen. Microbiol.*, 3, 85, 1976.
10. **Clarke, P. H. and Meadow, P. M.**, Evidence for the occurrence of permeases for tricarboxylic acid cycle intermediates in *Pseudomonas aeruginosa, J. Gen. Microbiol.*, 20, 144, 1959.
11. **Hamilton, W. A. and Dawes, E. A.**, The nature of the diauxic effect with glucose and organic acids in *Pseudomonas aeruginosa, Biochem. J.*, 76, 70P, 1968.
12. **Hamilton, W. A. and Dawes, E. A.**, Further observations on the nature of the diauxic effect with *Pseuomonas aeruginosa, Biochem. J.*, 79, 25P, 1961.
13. **Hamlin, B. T., Ng, F. M-W., and Dawes, E. A.**, Regulation'of enzymes of glucose metabolism in *Pseudomonas aeruginosa* by citrate, in *Microbial Physiology and Continuous Culture*, Powell, E. O., Evans, C. G. T., Strange, R. E., and Tempest, D. W., Eds., Her Majesty's Stationery Office, London, 1967, 211.
14. **Hamilton, W. A. and Dawes, E. A.**, A diauxic effect with *Pseudomonas aeruginosa, Biochem. J.*, 71, 25P, 1959.
15. **Ng, F. M-W. and Dawes, E. A.**, Chemostat studies on the regulation of glucose metabolism in *Pseudomonas aeruginosa* by citrate, *Biochem. J.*, 132, 129, 1973.
16. **Midgley, M. and Dawes, E. A.**, The regulation of transport of glucose and methyl α-glucoside in *Pseudomonas aeruginosa, Biochem. J.*, 132, 141, 1973.
17. **Campbell, J. J. R., Hogg, L. A., and Strasdine, G. A.**, Enzyme distribution in *Pseudomonas aeruginosa, Biochem. J.*, 83, 1155, 1962.
18. **Whiting, P. H., Midgley, M., and Dawes, E. A.**, The regulation of transport of glucose, gluconate and 2-oxogluconate and of glucose catabolism in *Pseudomonas aeruginosa, Biochem. J.*, 154, 659, 1976.
19. **Whiting, P. H., Midgley, M., and Dawes, E. A.**, The role of glucose limitation in the regulation of the transport of glucose, gluconate and 2-oxogluconate, and of glucose metabolism in *Pseudomonas aeruginosa, J. Gen. Microbiol.*, 92, 304, 1976.
20. **Tempest, D. W. and Neijssel, O. M.**, Microbial adaptation to low-nutrient environments, in *Continuous Culture 6: Applications and New Fields*, Ellis Horwood Ltd., Chichester, 1976, chap. 22.
21. **Dawes, E. A. and Senior, P. J.**, The role and regulation of energy reserve polymers in micro-organisms, *Adv. Microb. Physiol.*, 10, 135, 1973.
22. **Antoine, A. D. and Tepper, B. S.**, Environmental control of glycogen and lipid content of *Mycobacterium phlei, J. Gen. Microbiol.*, 55, 217, 1969.
23. **Stanier, R. Y., Doudoroff, M., Kunisawa, R., and Contopoulou, R.**, The role of organic substrates in bacterial photosynthesis, *Proc. Natl. Acad. Sci., U.S.A.*, 45, 1246, 1959.
24. **Wilkinson, J. F. and Munro, A. L. S.**, in *Microbial Physiology and Continuous Culture*, Powell, E. O., Evans, C. G. T., Strange, R. E., and Tempest, D. W., Eds., Her Majesty's Stationery Office, London, 1967, 173.
25. **Duguid, J. P., Smith, I. W., and Wilkinson, J. F.**, Volutin production in *Bacterium aerogenes* due to development of an acid reaction, *J. Pathol. Bacteriol.*, 67, 289, 1954.
26. **Herbert, D.**, The chemical composition of micro-organisms as a function of their environment, in *Microbial Reaction to Environment*, Meynell, G. G. and Gooder, H., Eds., University Press, Cambridge, 1961, 391.

27. **Holme, T.,** Continuous culture studies on glycogen synthesis in *Escherichia coli* B, *Acta Chem. Scand.,* 11, 763, 1957.
28. **Holme, T., Laurent, T., and Palmstierna, H.,** On the glycogen in *Escherichia coli* B; Variations in molecular weight during growth. II., *Acta Chem. Scand.,* 12, 1559, 1958.
29. **Holme, T.,** On the glycogen in *Escherichia coli* B; Variation in molecular weight during growth. III, *Acta Chem. Scand.,* 12, 1564, 1958.
30. **Schuster, E. and Schlegel, H. G.,** Chemolithotrophes wachstum von Hydrogenomonas H16 im chemostaten mit elektrolytischer knallgaserzeugung, *Arch. Mikrobiol.,* 58, 380, 1967.
31. **Schlegel, H. G., Gottschalk, G., and von Bartha, R.,** Formation and utilization of poly-β-hydroxybutyric acid by Knallgas bacteria *(Hydrogenomonas),* *Nature (London),* 191, 463, 1961.
32. **Stockdale, H., Ribbons, D. W., and Dawes, E. A.,** Occurrence of poly-β-hydroxybutyrate in the Azotobacteriaceae, *J. Bacteriol.,* 95, 1798, 1968.
33. **Senior, P. J., Beech, G. A., Ritchie, G. A. F., and Dawes, E. A.,** The role of oxygen limitation in the formation of poly-β-hydroxybutyrate during batch and continuous culture of *Azotobacter beijerinckii, Biochem. J.,* 128, 1193, 1972.
34. **Herbert, D., Phipps, P. J., and Tempest, D. W.,** The chemostat: design and instrumentation, *Lab. Pract.,* 14, 1150, 1965.
35. **Mackereth, F. J. H.,** An improved galvanic cell for determination of oxygen concentrations in fluids, *J. Sci. Instrum.,* 41, 38, 1964.
36. **Maclennan, D. G. and Pirt, S. J.,** Automatic control of dissolved oxygen concentration in stirred microbial cultures, *J. Gen. Microbiol.,* 45, 289, 1966.
37. **Jacob, H-E.,** Redox potential in *Methods in Microbiology,* Vol. 2, Norris, J. R. and Ribbons, D. W., Eds., Academic Press, London, 1970, 91.
38. **Ward, A. C., Rowley, B. I., and Dawes, E. A.,** Effect of oxygen and nitrogen limitation on poly-β-hydroxybutyrate biosynthesis in ammonium-grown *Azotobacter beijerinckii, J. Gen. Microbiol.,* 102, 61, 1977.
39. **Senior, P. J. and Dawes, E. A.,** The regulation of poly-β-hydroxybutyrate metabolism in *Azotobacter beijerinckii, Biochem. J.,* 134, 225, 1973.
40. **Jackson, F. A. and Dawes, E. A.,** Regulation of the tricarboxylic acid cycle and poly-β-hydroxybutyrate metabolism in *Azotobacter beijerinckii* grown under nitrogen or oxygen limitation, *J. Gen. Microbiol.,* 97, 303, 1976.
41. **Senior, P. J. and Dawes, E. A.,** Poly-β-hydroxybutyrate biosynthesis and the regulation of glucose metabolism in *Azotobacter beijerinckii, Biochem. J.,* 125, 55, 1971.
42. **Ritchie, G. A. F., Senior, P. J., and Dawes, E. A.,** The purification and characterization of acetoacetyl-coenzyme A reductase from *Azotobacter beijerinckii, Biochem. J.,* 121, 309, 1971.
43. **Parker, C. A.,** Effect of oxygen on nitrogen fixation by Azotobacter, *Nature (London),* 173, 780, 1954.
44. **Dalton, H. and Postgate, J. R.,** Effect of oxygen on growth of *Azotobacter chroococcum* in batch and continuous cultures, *J. Gen. Microbiol.,* 54, 463, 1969.
45. **Amarasingham, C. R. and Davis, B. D.,** Regulation of α-ketoglutarate dehydrogenase formation in *Escherichia coli, J. Biol. Chem.,* 240, 3664, 1965.
46. **Haaker, H. and Veeger, C.,** Regulation of respiration and nitrogen fixation in different types of *Azotobacter vinelandii, Eur. J. Biochem.,* 63, 499, 1976.
47. **Nagai, S., Nishizawa, Y., Onodera, M., and Aiba, S.,** Effect of dissolved oxygen on growth yield and aldolase activity in chemostat culture of *Azotobacter vinelandii, J. Gen. Microbiol.,* 66, 197, 1971.
48. **Nagai, S., Nishizawa, Y., and Aiba, S.,** Some consideration on the rate of induced aldolase synthesis in *Azotobacter vinelandii, J. Gen. Appl. Microbiol.,* 20, 229, 1974.
49. **Carter, I. S. and Dawes, E. A.,** Effect of oxygen concentration and growth rate on glucose metabolism, poly-β-hydroxybutyrate biosynthesis and respiration, in *Azobacter beijerinckii, J. Gen. Microbiol.,* 110, 393, 1979.

Chapter 2

BACTERIAL ENVELOPE STRUCTURE AND MACROMOLECULAR COMPOSITION*

D.C. Ellwood and A. Robinson

TABLE OF CONTENTS

---

* This chapter was submitted in May 1978.

I. INTRODUCTION

Clearly, in order to grow, microorganisms require a source of energy and supply of essential elements. These are C, N, O, H, S, P, Mg, K, and a variety of "trace" elements. It is possible to arrange the composition of the growth medium so that any of these elements is the growth limiting substrate when the organisms are grown in a chemostat.[1] When the growth of an organism is limited by a particular nutrient, then it is a prerequisite of chemostat theory that the organism will contain the minimum amount of that nutrient commensurate with the efficient functioning of the organism. Thus it is possible to grow different organisms under identical conditions in a chemostat and examine their composition in a comparative way. This kind of information will then allow inferences to be drawn as to the functioning of a given nutrient in different kinds of organisms.

These ideas were first exploited by Tempest and colleagues in exploring the role of magnesium, phosphorus, and potassium in Gram-positive and Gram-negative strains.[2] They grew the Gram-negative organism *Klebsiella aerogenes* in a chemostat with limiting magnesium, glycerol, or potassium at different dilution rates and at different temperatures. They observed that when the organism was grown magnesium limited the concentration of biomass present in the culture fell as the dilution rate (the growth rate) increased. This suggested that the cells' minimum requirement for magnesium increased as the dilution rate increased. It was known from the classical earlier work of Herbert[3] that the RNA content of *K. aerogenes* also increased markedly with faster growth rates of this organism in a chemostat (see Section IV.B). This led Tempest to compare the RNA-nucleotide:magnesium ratio for the organism grown magnesium-limited at different dilution rates. At all dilution rates studied (0.1 to 0.8 hr^{-1}) the mole ratio was almost constant. They concluded on this basis and on the evidence obtained by growing the organism at different temperatures at constant growth rates that in *K. aerogenes* the bulk of the cell-bound magnesium is located in some functional way with the ribosome.

When the Gram-positive organism *Bacillus subtilis* var. *niger* was examined in a similar way almost identical results were obtained. Thus it is this ribosomal requirement that demands the presence of a high concentration of magnesium in the growing cells.

This approach was extended to study the role of potassium in growing cells. It was found that with cultures of *K. aerogenes* growing potassium-limited, the potassium content of the organisms varied with the ribosome content, in the same way that the magnesium content had done. Indeed the ratio of cell-bound magnesium:potassium:RNA was roughly constant at 1:4:5 for all growth rates studied. This then suggested a relationship between ribosome function and the magenesium and potassium requirements of the Gram-negative organism.

However when the Gram-positive organism *B. subtilis* var. *niger* was grown potassium-limited and its composition compared with that of *K. aerogenes* grown under identical conditions marked differences were found. The content of magnesium and RNA of these two organisms were similar at different dilution rates, but the content of potassium and phosphorus were markedly different. The Gram-positive organism contained about three times as much potassium and 50% more phosporus than the Gram-negative organism. Thus it seems that *B. subtilis* var. *niger* has an excess requirement for phosphorus and potassium when compared to the Gram-negative organism, (Table 1). It seems unlikely that this was related to ribosomal function and the "excess" phosphorus was further investigated. (see Section III.A.1).

These results show the value of the approach of growing an organism in conditions that allow the minimum amount of an essential nutrient to be present in a cell. This

Table 1
INFLUENCE OF DILUTION RATE ON
THE POTASSIUM, MAGNESIUM,
PHOSPHORUS, AND RNA CONTENTS
OF *KLEBSIELLA AEROGENES* AND
BACILLUS SUBTILIS VAR.
NIGER GROWING IN K^+-LIMITED
CHEMOSTAT CULTURES AT A FIXED
TEMPERATURE (35°) AND pH VALUE
(6.7)

Dilution rate (hr^{-1})	g/100 g dry weight bacteria			
	Potassium	Magnesium	Phosphorus	RNA
Baccillus subtilis var. *niger*				
0.2	3.7	0.18	3.2	12.8
0.4	4.9	0.22	3.5	14.0
Kiebsiella aerogenes				
0.2	1.08	0.16	1.7	11.7
0.4	1.42	0.21	2.2	15.0

From Tempest, D. W., *Adv. Microb. Physiol.*, 4, 223, 1970, Academic Press, London. With permission.

minimum amount will operate at maximum efficiency giving the microbial physiologist an invaluable technique to elucidate the involvement of the limiting nutrient in the "economy" of the cell. The results of studies on bacterial macromolecular (in particular envelope) structure are described in the following sections.

II. MORPHOLOGY

It is generally found that bacteria growing in a chemostat vary in size as a function of growth rate for any specific growth limiting nutrient. Organisms grown at a slow rate are smaller than those grown at a faster rate, thus these cells would have a greater surface to volume ratio. If surface area is related to the uptake of nutrients then the uptake systems would be greater per unit mass for smaller organisms.

This smaller size of organism at slower growth rates is reflected in the amount of cell envelope present as a percentage of cell mass for a number of organisms. The organisms that show this reponse of varied envelope content are usually of the bacillary shape (Bacillus, Klebsiella, Enterobacter), while those of coccal form (staphylococci, streptococci) show little difference in envelope content with varying growth rate.[4]

Arthrobacter strains change their morphology during growth from rods to cocci. In the early stages of batch culture, rods predominate whereas cocci forms are characteristic of the late lag and stationary phases.[5] This system was examined by Luscombe and Gray[6,7] who showed that the change in morphology from rods to cocci occurred at a characteristic growth rate of the organism when grown in a chemostat. This specific growth-rate-dependent change occurred when either carbon or nitrogen-limiting conditions were used. The chemical changes relating to this change in morphology are not yet clear, but it is interesting to note that the phosphorus content more than doubles in walls of cocci as compared to walls of rods.

The organism *Streptococcus mutans* has been implicated as the etiologic agent of dental caries and has been subject to a number of chemostat studies. When the organism is grown glucose-limited in a complex medium or in a chemically defined medium, there are interesting morphological changes with different growth rates.[8] At low growth rates (D = 0.05 hr^{-1}) *S. mutans* grows as single organisms with some short chains. As the growth rate is increased, the organisms occur as longer chains with fewer single bacteria and the chains seem to bind together to form agglomerates of large numbers of bacteria (200 to 300 per agglomerate). This effect is most marked at μ_{max} (in this system about a D = 0.5 hr^{-1}) when the whole culture ocurs in these large agglomerates. At this growth the organisms are very sticky and wall growth blocking the overflow on the chemostat becomes a serious problem. These results are of interest in that the ability to adhere is of course one of the virulence characterics of this organism.

This interesting behavior also occurs when sucrose is used to limit the growth of *S. mutans*, but in this case when the dilution rate is increased to D = 0.3 hr^{-1} the agglomerates take on a harder appearance and extracellular material adhering to the cells can be seen by electron microscopy. Cultures of this organism grown in glucose excess with amino acids, magnesium, or phosphorus as the limiting substrate do not show this behavior. However cultures of this organism grown phosphorus limited grow as long chains and the morphology is more characteristic of the classical picture of streptococci found in medical microbiological textbooks.

III. ENVELOPE STRUCTURE

A. Envelopes of Gram-positive Bacteria

Bacteria can be regarded as biochemical factories surrounded by a membrane and wall (together termed the envelope), which hold the cell together and allow the entry of nutrients into the cell and exit of waste products from it. Bacterial envelopes can be isolated after disruption of the organisms using standard techniques such as vigorous shaking with small diameter glass beads. The Gram-stain is a simple chemical staining method by which organisms may be classified. The envelope structure of bacteria can be directly correlated with their reaction to the Gram-stain.

Walls of Gram-positive bacteria are of relatively simple composition. They are composed principally of peptidoglycan, which takes up the shape of the bacteria holding the cytoplasm together. If synthesis of peptidoglycan is inhibited by, for example, penicillin the cytoplasm is no longer contained and the cell "bursts". Peptidoglycan can account for up to 90% of the dry weight of the Gram-positive cell wall; anionic polymers (teichoic acids and teichuronic acids), polysaccharides and proteins are also found in the walls. Proteins of Gram-positive cell walls have been little studied. In contrast to Gram-negative bacteria the cell wall of Gram-positive bacteria can be readily separated from the cytoplasmic membrane.

1. Anionic Polymers
a. Teichoic Acid

Teichoic acids are a group of polymers containing phosphate that have primarily been isolated from walls and membranes of Gram-positive bacteria. The first types of these polymers to be described were isolated in 1957 from strains of *Lactobacillus arabinosus* and *Bacillus subtilis* following the identification of cytidine diphosphate glycerol and cytidine diphosphate ribitol in these organisms.[9] The polymers were composed of ribitol phosphate and glycerol phosphate units. Since then there has been considerable investigation of these polymers and the term teichoic acid is now used to include all bacterial wall, membrane, or capsular polymers containing glycerol phosphate or ribitol phosphate residues. Even with this broad definition there are phosphate-containing polymers that fall outside it. These are the sugar-1-phosphate or oligosaccharide-1-phosphate polymers found in walls of some bacteria. Conveniently though, they can be considered with teichoic acids.

In general the teichoic acids have been isolated exclusively from Gram-positive bacteria. With a few exceptions Gram-positive bacteria contain a glycerol phosphate based membrane teichoic acid. Walls of a number of Gram-positive bacteria contain a teichoic acid and this may account for up to 70% by weight of the wall when the organisms are grown magnesium-limited. The capsular substance of some sero-types of the Gam positive organism *Diplococcus pneumoniae* are complex teichoic acids. [10] More recently the capsular substance of the Gram-negative organism *Haemophilus influenzae* has been shown to be a sugar ribitol phosphate diester polymers.[11] This is the first real evidence of a teichoic acid in Gram-negative bacteria, but as the *Escherichia coli* K100 antigen cross-reacts with these *H. influenzae* polymers the K100 antigen may also turn out to be a teichoic acid.[12] The structures of several teichoic acids and related polymers are as follows:

(i)

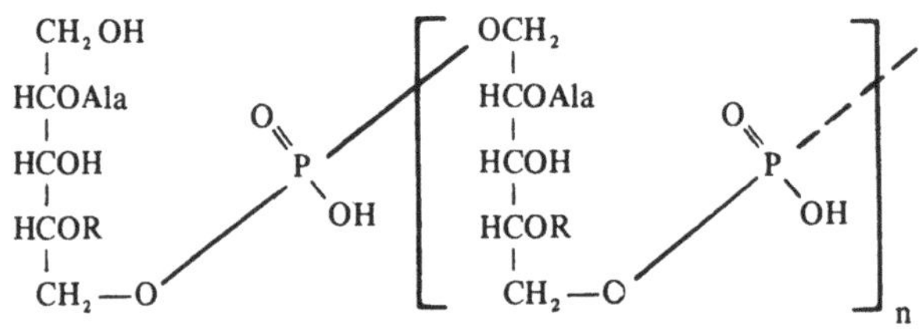

R = α or β-N-acetylglucosaminyl
Ala = D-alanine

STRUCTURE (1)

R = α or β-N-acetylglucosaminyl
Ala = D-alanine

Ribitol teichoic acid from *Staphylococcus aureus*

(ii)

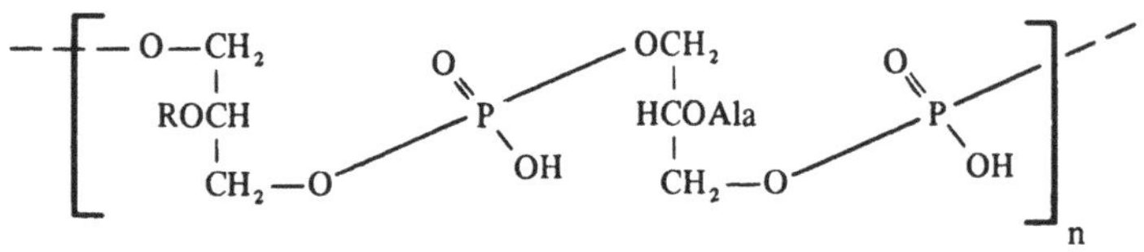

R = H or glycosyl

STRUCTURE (2)

R = H or glycosyl
A glycerol teichoic acid

(iii)

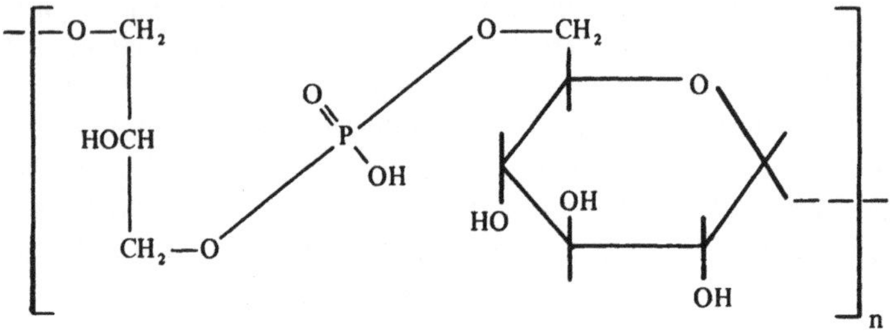

STRUCTURE (3)

Glucosylglycerol phosphate teichoic acid from *Bacillus licheniformis* ATCC 9945

(iv)

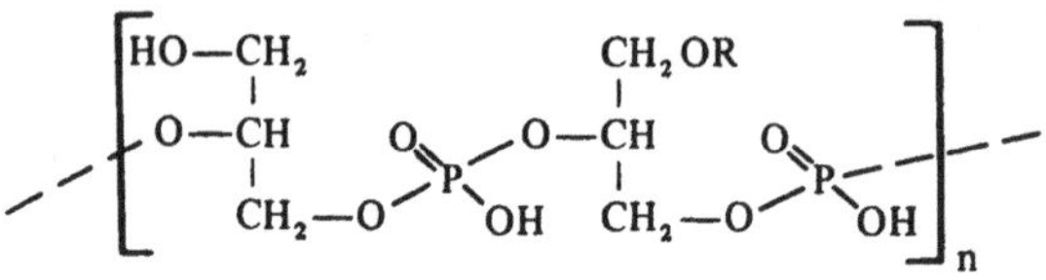

R = glucosyl

STRUCTURE (4)

R = glycosyl
A glycerol 1:2 teichoic acid from *Bacillus subtilus* var. *niger*.

(v)

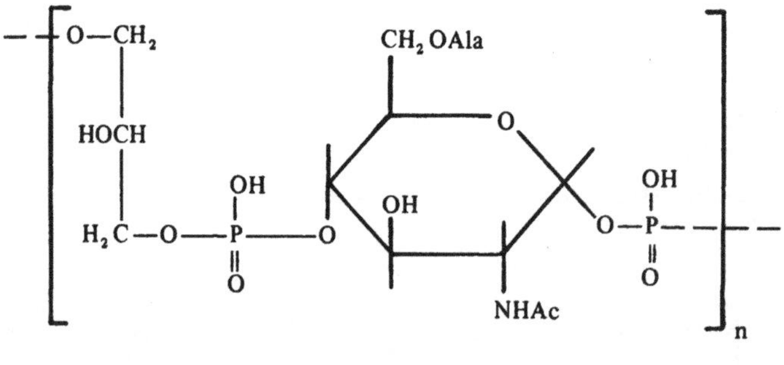

STRUCTURE (5)

Teichoic acid containing sugar-1-PO_4 linkages from *Staphylococcus lactis* 13

b. Teichuronic Acid

Teichuronic acids are also anionic polymers, but lacking phosphate residues. They contain acidic sugars as part of the polysaccharide chain. They occur in the walls of some Bacillus strains when grown in batch cultures and have the general structure as follows:

Glucuronic acid   N-acetyl glucosamine

STRUCTURE (6)

Teichuronic acid from Bacillus licheniformis

Such acidic sugar-containing polysaccharides have been identified as being produced by a number of strains of bacilli and staphylococci probably as extracellular polysaccharides. It is also interesting to note that a number of extracellular polysaccharides produced by Gram-negative bacteria contain acidic side chains in the polysaccharide, e.g., *Xanthomonas campestris.*

c. Effect of Nutrient Limitation on Wall Teichoic/Teichuronic Acid Content

As we stressed earlier in this chapter, it is a precept that bacteria grown in a chemostat with a limiting substrate utilize that substrate to the maximum. It was observed that *B. subtilis* var. *niger* grown phosphate limited contained less phosphate than when grown in the presence of excess phosphate. This observation led to the question of how and why this had occurred, and to a fundamental investigation of the effect of growth environment on the composition of cell walls.[4]

The organism studied initially was *B. subtilis* var. *niger* and this organism was grown in a chemostat with a variety of different limiting substrates. The wall content of cells and the compositions of the isolated walls were established (Table 2). When cultures were grown under conditions of phosphate excess, the wall invariably contained phosphate. This phosphate was shown to be due to the presence of a glycerol teichoic acid by the usual methods (see later). However, when the organism was grown phosphate-limited the phosphorus in the wall fell to a low value. No teichoic acid could be detected in the wall, but teichuronic acid was found to be a major wall polymer. It seemed that when a constraint was applied to the supply of phosphate, the synthesis of wall teichoic acid was switched off. However an anionic polymer was still required and hence the phosphorus-free teichuronic acid was synthesized.

An examination of Table 2 shows that the limitation that produces organisms having the largest amount of teichoic acid in the wall is magnesium. Again arguing on the basis that an organism grown nutrient limited uses the limited nutrient most efficiently, this result suggests that teichoic acid in the wall is involved in the uptake of magnesium. This strongly supports the earlier suggestion of the involvement of teichoic acid in cation uptake.[13] Further evidence to support this idea on the basis of results obtained on the uptake of magnesium by isolated wall preparations of several bacteria has been obtained by Baddiley and co-workers using X-ray photoelectron techniques.[14]

A number of organisms have now been grown in a chemostat either magnesium- or phosphate limited, and their walls examined for their phosphorus contents. (Table 3). The strains of bacilli examined have all been grown in a simple salts medium and they all show this loss of teichoic acid when grown phosphate limited. Archibald and co-

Table 2

EFFECT OF DIFFERENT GROWTH LIMITATIONS ON THE CELL-WALL CONTENT AND COMPOSITION OF *BACILLUS SUBTILIS* VAR. *NIGER* GROWN IN A CHEMOSTAT CULTURE WITH THE DILUTION RATE FIXED AT 0.3 HR⁻¹, THE TEMPERATURE AT 35° AND pH7.0

Growth-limiting nutrient	Cell wall content (% dry weight)		Content (in g/100 g dry weight isolated cell walls) of					
			Protein	Phosphorus	Teichoic acid[a]	Hexose	Glucuronic acid	Galactosamine
Glucose	12.0;	13.2	10.0	3.9	35-42	20	<3	<3
Ammonia	21.0;	21.0	11.5	4.8	45-52	23	<3	<3
Sulphate	21.0;	21.0	12.5	4.9	45-53	23	<3	<3
K⁺	15.3;	15.5	10.5	4.3	39-46	20	<3	<3
Mg²⁺	17.8;	18.0	12.5	6.9	62-74	32	<3	<3
Phosphate	17.0;	17.5	10.0	0.5	<3	<2	25	17

[a]   Assuming the teichoic acid to be fully glucosylated, but without alanine.

From Ellwood, D. C. and Tempest, D. W., *Adv. Microb. Physiol.*, 7, 83, 1972, Academic Press, London, with permission.

Table 3

COMPARISON OF PHOSPHORUS CONTENTS OF WALLS FROM DIFFERENT GRAM-POSITIVE BACTERIA GROWN IN Mg²⁺-LIMITED OR PO₄³⁻ -LIMITED CHEMOSTAT CULTURES.

Organism	Gram phosphorus/100 g dried walls	
	Mg²⁺-limitation	PO₄³⁻ - limitation
Bacillus subtilis W23*	4.1	0.1
Bacillus subtilis 168*	4.2	0.1
*Bacillus subtilis**	3.9	0.1
Bacillus subtilis var. *niger**	6.0	0.4
Bacillus licheniformis NCTC 6346*	3.6	0.2
Bacillus megaterium KM*	0.8	0.04
Staphylococcus aureus H*	3.0	0.5
Staphylococcus epidermidis NCTC 2102	3.0	0.1
Staphylococcus albus NCTC 7944	2.1	0.4
Streptococcus faecalis NCIB 8191	1.5	0.5
Streptococcus mutans BHT	1.5	0.7
Lactobacillus arabinosus	2.4	2.1

*   Strain of *Bacillus subtilis* obtained from Prof. J. Baddiley.

*   From Ellwood, D. C. and Tempest, D. W., *Adv. Microb. Physiol.*, 7, 83, 1973, Academic Press, London. With permission.

workers have also observed the loss of teichoic acid when *B. subtilis* W23 was grown phosphate limited.[15]

It is much more difficult to carry out similar experiments with organisms that require a complex medium. In the experiments carried out at Porton, we have used precipitation methods to remove phosphate and magnesium from complex medium and then added back the required amount of the limiting component. Using this technique we were able to show a large drop in the wall phosphorus content of a number of staphylococci and streptococci grown phosphate limited compared with organisms grown magnesium limited. However Archibald et al. were unable to obtain any effect on the wall phosphorus content of a strain of Staphylococci grown phosphate limited in a chemically defined amino acid medium.[16] We obtained a similar result when we grew *Lactobacillus arabinosus* (the wall of which contains teichoic acid when grown in batch culture) in a complex chemically defined medium. It would be of considerable interest to extend these studies to see how many teichoic acid-containing bacteria undergo this teichoic acid/teichuronic changeover.

The phosphoglucomutase-deficient mutants of *Bacillus licheniformis* are unable to handle glucose and hence are not able to form teichuronic acid or to form a glycosylated teichoic acid when grown in batch culture. When these mutants were examined in a chemostat, they were able to form a nonglucosylated glycerol teichoic acid when grown magnesium limited. However when grown phosphate limited they responded by, as expected, not forming teichoic acid, but as they were unable to make the expected teichuronic acid the amount of mucopeptide in the wall increased markedly. These changes were also reflected in some interesting morphological variations in the cells.[17] In all of these experiments some residual phosphate was found in the walls of phosphate-limited organisms. This could be accounted for by an absolute requirement for some phosphate-containing compound in the wall. In this regard it is interesting to note that Heckels et al. showed that short chains of glycerol phosphate units joined the mucopeptide to the ribitol teichoic acid in staphylococci.[18] If this type of unit also attached teichuronic acid residues to the mucopeptide it could give rise to the absolute requirement of some phosphorus in the wall.

An alternative explanation is that this phosphorus found in wall preparations could arise from contamination with lipoteichoic acid (i.e., membrane associated teichoic acid). Lipoteichoic acid was found to be present in *B. subtilis* W23 even when the organism was grown phosphate limited, suggesting a specific functional role for these compounds. These compounds are discussed below.

d. Lipoteichoic Acid

In most of the Gram-positive bacteria so far examined, there is found a glycerol teichoic acid fraction that can be extracted from the cell contents with trichloroacetic acid or phenol. When protoplasts are made in the absence of any Mg^{2+} ions, the teichoic acid is released from the protoplasts suggesting that this teichoic acid lies between the wall and the membrane.

This behavior became clear when Wicken and Knox showed that this teichoic acid had a lipid tail and they called this class of teichoic acid "Lipoteichoic acids."[19] It appears that these teichoic acids are composed of long chains of the classical 1-3 linked glycerol phosphate units attached via sugar residues to a fatty acid acylated glycerol moiety as follows:

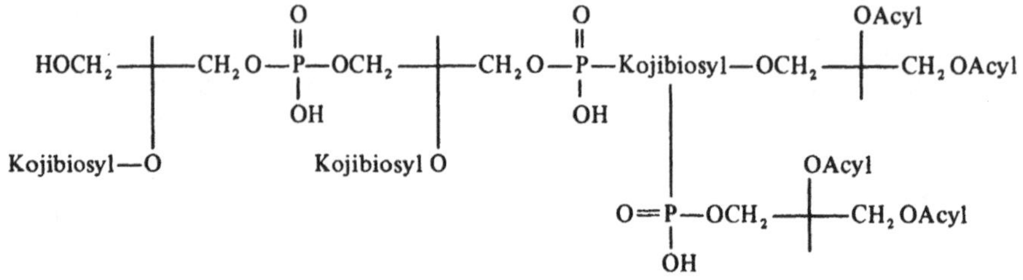

STRUCTURE (7)

Structure of the lipoteichoic acid of *Streptococcus faecalis*
Acyl = fatty acids

It is thought that the lipid tail anchors the teichoic acid in the membrane. Furthermore Wicken and Knox have shown that the lipoteichoic acid is excreted into the medium. This excretion continues even when the organism *(Streptococcus mutans* BHT) is grown in a chemostat under energy- (glucose) limitation.[20] It is surprising that such excretion should occur under these conditions, and tends to suggest that this excretion has some functional significance.

e. Function of Anionic Polymers

It seems that a major role of teichoic acids may be in the uptake of cations, and possibly in the control of cell lytic activities during growth. They also have important serological properties, which of course may not be important to the cell itself, but are of increasing influence on our ideas with respect to the pathogenicity of Gram-positive bacteria, e.g., in dental caries.

f. Turnover of Walls

One of the most powerful applications of the chemostat technique is to study the kinetics of changeover from one limitation to another. Clearly in the above cell wall studies this marked change in cell wall composition could have been due to the selection of mutants (it is known that the chemostat is fiercely selective for mutants) and not due to any phenotypic change. Thus when a chemostat culture of *B. subtilus* vas *niger* was changed from being magnesium limited to being phosphate limited, the wall teichoic acid was depleted at a faster rate than that predicted for the washout rate.[4] Furthermore the wall teichuronic acid increased substantially faster than that suggested by the growth rate. (Figure 1). These rates were very much faster than mutant selection would allow, and it was clear that a phenotypic change was occurring. Furthermore because the changes were faster than simple washout would allow, wall turnover, i.e., active extrusion of teichoic acid, occurred.

These results led Glaser and group to examine wall turnover of bacilli strains in batch experiments. They found that the turnover rates of mucopeptide and teichoic acid are identical.[21] Since then a number of Gram-positive bacteria including streptococci and lactobacilli strains have been shown to undergo wall turnover.

This phenomenon of cell wall turnover during growth in a chemostat has been very elegantly studied by Archibald and co-workers.[22] *Bacillus subtilis* W23 contains a ribitol teichoic acid in its wall and this is the binding site for phage SP50. These workers grew the organism in the chemostat potassium limited so that it contained teichoic acid in its wall, and by electron microscopy they showed that the phage bound to the cell surface all over the cell. When the growth-limiting nutrient was changed to phosphorus, the changeover from teichoic acid to teichuronic acid was visualized by studying

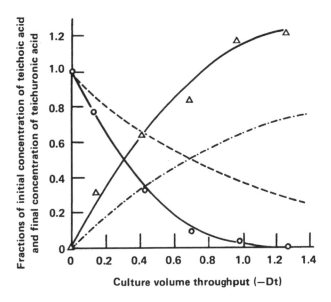

FIGURE 1.   Changes in the teichoic acid and teichuronic acid contents of cell walls of *B. subtilis* var. *niger* following changeover from conditions of magnesium-limitation to phosphate-limitation. The regular broken line represents the theoretical washout rate assuming teichoic acid synthesis ceased immediately the environment became depleted of phosphate. The irregular broken line represents the theoretical rate of increase in teichuronic acid assuming that its synthesis started immediately the culture became phosphate-limited and continued at a rate proportional to the rate of synthesis of biomass. (O) Actual teichoic acid content and (Δ) actual teichuronic acid content. (From Ellwood, D. C. and Tempest, D. W., *Adv. Microb. Physiol.*, 7, 83, 1972. With permission.)

the phage absorption in the electron microscope. The results showed that during the changeover cells were found with phage predominantly binding to the ends of the bacteria. This indicated a very marked spatial turnover of the wall.

It seems very clear from all these studies that the wall is in a more dynamic state than hitherto supposed. This concept has profound implications on our thinking in respect to the behavior of organisms in nature.

g. Effects of Growth Rate

It was known from batch studies that the teichoic acid content of walls of bacilli varied quantitatively throughout the so-called growth cycle, being greater in midlog phase cells than in those in the late stationary phase.[23,24]

In chemostat cultures of *B. subtilis* var. *niger*, the content of teichoic acid in the wall increased as the dilution rate was raised in magnesium-limited cultures (50% at D = 0.1 hr^{-1} to 75% at D = 0.3 hr^{-1}).[4] When this organism was grown phosphate limited, the content of teichuronic acid in the wall also increased, though not markedly, as the dilution rate was raised. Interestingly teichoic acid was not found in the walls of organisms until the growth rate was approaching μ_{max} (in this system D = 0.6 hr^{-1}). This is in contrast to the data of Wright and Heckels[25] who found that with *B. subtilis* W23 teichoic acid started to occur in walls of cells grown phosphate limited at a D = 0.3 hr^{-1}. In recent experiments the rate of wall turnover has been shown to be related to growth rate in *Bacillus subtilis* W23.[26]

h. Effect of Ionic Environment and pH

Teichoic acids form a densely charged environment in or beneath the wall, and are involved in concentrating the available cations for cells. These anionic polymers would appear to be more effective at concentrating magnesium ions for assimilation than teichuronic acid, because it has been shown that walls from magnesium-limited organisms bind magnesium ions more avidly than walls from phosphate-limited organisms.[4] If a constraint is applied to magnesium uptake, for example by growing the organism in a medium containg 6% (w/v) NaCl, high affinity magnesium binding walls have to be produced. This is achieved in *B. subtilis* var *niger* when grown under phosphate limitation in the presence of 6% (w/v) NaCl by production of teichoic acid in the walls and not the expected teichuronic acid.[4] Thus teichuronic acids appears to be able to functionally replace teichoic acid under conditions of magnesium excess provided there is no constraint on magnesium uptake.

Varying the culture pH, not surprisingly in view of the proposed function of these anionic polymers, alters the composition and content of teichoic acid in walls.[4] The teichoic acid content of magnesium-limited *B. subtilis* W23 decreased when the culture pH was decreased from 7.5 to 5.5, and there was a detectable increase in teichuronic acid. Probably more significantly as the pH was lowered there was an increase in the amount of ester-bound alanine in the teichoic acid. The alanylation appears to be essential for efficient functioning at low pH, because it occurs in the teichoic acid of *B. subtilis* var *niger* when grown under nitrogen limitation. Lowering the pH of phosphate-limited *B. subtilis* var. *niger* reduced the teichuronic acid content and caused a marked increase in teichoic acid, not normally found in phosphate-limited organisms, and again this teichoic acid was alanylated

2. Peptidoglycan

The peptidoglycan (otherwise termed murein, glycopeptide, or mucopeptide) component of bacterial envelopes is of great importance as the main supporting medium for the envelope and also in shape maintenance. (The question of shape determination in bacteria has been discussed in some detail by Henning [27]). Peptidoglycans are built up to form a close network of identical chains and form large bag-shaped macromolecules (the 'sacculus') enclosing and protecting the cytoplasmic membrane. The composition and primary structure of peptidoglycan have been extensively studied in the last two decades. The basic structure of the peptidoglycan in almost all bacterial envelopes is a backbone of alternating β-1:4 linked N-acetylglucosamine and N-acetylmuramic acid molecules. The peptide moiety of the peptidoglycan is bound through its N-terminus to the carboxyl group of muramic acid and contains L- and D-amino acids. The peptide cross links are seemingly more variable than the glycan chain and differences in the peptide structure have been considered for the classification of Gram-positive bacteria.[28]

Changes in the structure of the peptidoglycan with growth environment do undoubtedly occur. The changes in structure due to batch growth on different media, in particular the effects of different amino acid limitation, have been extensively reviewed by Schliefer et al.[29] The effects of changes in batch culture were not as pronounced as those resulting from growth of the organisms in different limited environments in the chemostat.[4,30] Principally *Bacillus subtilis* var *niger* and *B. subtilis* W23 have been used for these studies. The lysozyme sensitivities of heated (100°C, 20 min, to inactivate autolytic enzymes) *B. subtilis* W23 from phosphate-, ammonia-, or magnesium-limited chemostat cultures were found to vary markedly (Figure 2). Magnesium-limited cells lysed slower than phosphate- or ammonia-limited cells. These differences could not be relatd to the teichoic acid or teichuronic acid contents of the cells. Quantitative analy-

51

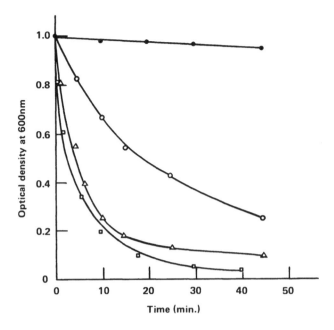

FIGURE 2.   Effect of lysozyme (20 μg/mℓ) on the optical density of
suspensions of *B. subtilis* W23. The organisms were grown at D =
0.3 h⁻¹, 35° and pH 7.0 under magnesium- (O), phosphate- (Δ), or
ammonium-limited (□) conditions. A control suspension of magnesium-limited organisms contained no lysozyme (•). (From Ellwood,
D. C. and Tempest, D. W., *Adv. Microb. Physiol.*, 7, 83, 1972. With
permission.)

ses of the constituents of the acid-hydrolysed isolated peptidoglycans from these organisms revealed substantial differences in the muramic acid and glucosamine contents. There was little variation in the molar proportions of alanine:glutamic
acid:diaminopimelic acid (Table 4). The molar ratios of muramic acid and glucosamine
to glutamic acid in the envelopes of ammonia-limited organisms (most sensitive to
lysozyme) were about unity, close to the theoretical values. With magnesium-limited
(and sulphate-limited)organisms, however, the peptidoglycan contained significantly
more glucosamine than muramic acid, whereas envelopes from potassium, glucose-,
or phosphate-limited *B. subtilis* W23 contained more muramic acid than glucosamine.
Variation in the temperature of growth of the organism also altered the peptidoglycan
composition. The muramic acid and glucosamine contents of isolated envelopes from
magnesium-limited organisms, D = 0.2 hr⁻¹, increased with decreasing temperature.
Lowering the growth temperature thus had a similar effect to increased growth rate
on the peptidoglycan structure.

Johnson and Campbell[31] in some elegant chemical studies examined the effect of
batch growth in defined synthetic or complex medium on the peptidoglycan of *Micrococcus sodonensis*. Essentially they found that envelopes from organisms grown in a
complex medium were more lysozyme resistant than envelopes from organisms grown
in the defined medium due to (1) an increased level of 0-acetyl substitution, and (2)
an increased level of the complexity of the cross linked structure. Archibald and Heckels[16] found that a nutritional mutant of *Staphylococcus aureus* H grown in a defined
potassium-limited medium in a chemostat possessed walls that were more susceptible
to the action of lytic enzymes than walls from batch growth cells. The continuously
grown cells had a less highly crosslinked peptidoglycan.

Table 4

COMPARISON OF PEPTIDOGLYCAN FRACTIONS FROM THE ENVELOPES OF
BACILLUS SUBTILIS W23 GROWN UNDER DIFFERENT CONDITIONS.

	Growth limiting substrate and dilution rate								
Component	Mg^{2+} 0.1 hr^{-1}	Mg^{2+} 0.2 hr^{-1}	Mg^{2+} 0.3 hr^{-1}	K^+ 0.3 hr^{-1}	Glucose 0.3 hr^{-1}	NH_3 0.3 hr^{-1}	SO_4^{2-} 0.3 hr^{-1}	PO_4^{2-} 0.3 hr^{-1}
Muramic acid	0.62	0.75	0.90	0.82	1.05	1.07	0.50	1.02
Glucosamine	0.79	0.92	1.04	0.67	0.70	1.03	0.70	0.70
Glutamic acid	1.00	1.00	1.00	1.00	1.00	1.00	1.00	1.00
Alanine	1.4	1.6	1.6	1.5	1.5	1.6	1.5	1.5
Diaminopimelic acid	0.93	1.04	1.07	0.93	1.15	1.15	1.18	1.17

Note: The amino compounds were analyzed using a Technicon TMS amino acid analyzer. The envelopes were hydrolyzed in 6N HCl for 18 hr after removal of teichoic acid and/or teichuronic acid with trichloracetic acid

From Ellwood, D. C., and Tempest, D. W., *Adv. Microb. Physiol.*, 7, 83, 1972, Academic Press, London. With permission.

Clearly subtle changes in the growth environment can bring about marked changes in the peptidoglycan structure, and again structural data from organisms grown in uncontrolled environments must be treated with some skepticism.

3. Cytoplasmic Membrane

Most of the work done on the phenotypic variability of the cytoplasmic membranes of Gram-positive organisms has involved the analysis of the lipid components, although alterations in the protein content do occur. Again much work has been done in batch culture especially on the effect of varying the temperature and pH of growth.[4,32,33] Temperature shifts affected the content and degree of saturation of the lipid fatty acid moieties, the lower the growth temperature the greater the proportion of unsaturated fatty acids in the bacterial lipids.

Ellwood and Tempest[4] examined the envelope composition of *B. subtilis* (designated *Bacillus polymyxa*) grown in various limited chemostat cultures. Sulfate-and magnesium-limited organisms had a high content of envelope lipid (39 and 25 %, respectively) whereas ammonia-limited organims had a low envelope lipid content (11%). Glucose- and phosphate-limited organisms had intermediate values. The lipid contents of the envelopes in all cases were higher than those from batch grown organisms (5%).

Not surprisingly phosphate limitation has a pronounced effect on the phospholipid compositions of the organisms (see also Section IIIB2). Minnikin and Abdolrahimza-deh[34] found significant changes in the relative proportions of individual polar lipids of two strains of *B. subtilis* when the pH of chemostat cultures was varied. In phosphate- and magnesium-limited cultures of *B. subtilis* var *niger* N.C.I.B. 8058 lysyl-phosphatidyl glycerol was present in higher proportions at pH 5.1 compared to organisms grown at neutral pH. Magnesium-limited organisms grown at pH 8.0 had no lysylphosphatidyl glycerol or phosphatidyl ethanolamine. Phosphate-limited *B. subtilis* N.C.I.B. 3610 had no phosphatidyl ethanolamine or lysylphosphatidyl glycerol when grown at neutral pH, but contained substantial proportions of these lipids when grown at low pH. Phosphatidyl glycerol was always lower in chemostat cultures than lysyl-phosphatidyl glycerol even at low pH when the lysylphosphatidyl glycerol content increased. The reverse is quite common in batch cultures. Production of the lysine derivative of phosphatidyl glycerol appears to be concerned with cation control at low pH and may be related to the increase in ester bound alanine in teichoic acids at low pH (see Section III.A.1.h).

Drucker et al.[35-38] have used the precise controlled growth environment of the chemostat to examine the fatty acid fingerprints of various streptococci using computer data analysis. The fatty acid composition of *Streptococcus mutans* was found to vary with the carbohydrate source and also oxygenation, which affected the degree of saturation of the fatty acids. Vitamin deficiency caused a pronounced effect on the fatty acid fingerprint of *Streptococcus* sp. SS with an almost complete loss of a major peak (iso-myristate). This work stresses the importance of rigidly controlling the growth conditions employed in chemotaxonomic work even with regard to trace substances in the medium.

B. Envelopes of Gram-negative Bacteria

The cell envelope of a Gram-negative bacterium is a complex structure consisting of at least three layers, the inner (cytoplasmic) membrane, a peptidoglycan layer, and an outer membrane.[39] Several techniques are available for the separation of the inner and outer membranes all based on the original work of Osborn et al.[40] The procedure involves separating the membranes, produced by lysing spheroplasts, on equilibrium sucrose density gradient centrifugation. This has considerably simplified the study of

Gram-negative envelopes. The inner membrane appears to be a fairly typical lipid/ protein membrane. The protein components have been shown to be extremely heterogeneous by sodium dodecyl sulphate-polyacrylamide gel electrophoresis (SDS-PAGE) and a vast array of enzymic activities have been assigned to the membrane (for example respiratory electron transport chain, and enzymes for the synthesis of envelope polymers). The peptidoglycan layer of Gram-negative bacteria has a similar structure to that of Gram-positive bacteria, but generally constitutes a lower proportion of the envelope and will not be considered separately here.

The outer membrane is a complex structure that accounts for the major differences between Gram-positive and Gram-negative bacteria. There are only a few major protein components, usually four,[41] in the membrane and they have all been isolated. The smallest of these proteins is a lipoprotein with a molecular weight of about 7000 daltons, which is covalently linked to the peptidoglycan.[42] Other proteins are also peptidoglycan associated.[43] Mutants lacking an outer membrane protein usually compensate for this by increasing the quantities of other proteins. However mutants of *E. coli* lacking all "major" outer membrane proteins have been isolated.[44] The outer membrane presents a formidable permeability barrier to the entry of molecules into the cell, and the outer membrane proteins have been implicated in pore formation. The lipopolysaccharide of the envelope is also a constituent of the outer membrane. The phenotypic structural variability of isolated lipopolysaccharide of organisms grown in continuous culture has been little studied. However studies of the biological variability of cell envelopes suggest that the lipopolysaccharide structures may vary phenotypically (see Section III.D).

1. Lipids

The lipid compositions of Gram-negative bacteria have been found to be markedly variable as have those of Gram-positive organisms (see Section III.A.3). Phosphate-limitation has a pronounced effect on the phospholipids of chemostat cultures of Gram- negative bacteria. For example Minnikin and Abdolrahimzadeh[45] found that magnesium-limited cultures of *Pseudomonas fluorescens* NCMB 129 have as major lipids phosphatidyl ethanolamine, phosphatidyl glycerol, and diphosphatidyl glycerol. Under conditions of phosphate limitation these phospholipids were replaced by an ornithine-amide lipid as almost the sole polar lipid component. The general structure of the ornithine-amide lipid is as follows:

$$H_2N \cdot (CH_2)_3 \cdot CH \cdot COOH$$
$$R_1 \cdot CH \cdot CH_2 \cdot CO \cdot NH \qquad R_1 \text{ and } R_2 = \text{alkyl groups}$$
$$R_2 \cdot CO \cdot O$$

STRUCTURE (8)

Magnesium-limited cultures of *Pseudomonas diminuta* have a similar lipid composition to that of batch cultures. Under phosphate limitation, however, the phospholipids (phosphatidyl glycerol and phosphatidyl glucosyl diglyceride) were almost totally replaced by acid glycolipids (glucosyl diglyceride, glucuronosyl-diglyceride and glucosylglucuronosyldiglyceride[46]). The phospholipids of *Pseudomonas aeruginosa* have also been found to be much reduced when the organism was subjected to phosphate limitation.[47] It would thus appear that phospholipids, once thought to be essential membrane components are indeed dispensible. The loss of phospholipids under phosphate limitation appears to be analogous to the teichoic acid/teichuronic acid changeover occurring due to phosphate limitation of Gram-positive bacteria (See Section III.A.1.c).

Gill[48] has used one of the attributes of the continuous culture in a study of the lipids of *Ps. fluorescens* in that he has studied the effect of temperature variation on the lipids of organisms grown at constant growth rates and vice versa, i.e., the temperature and rate of growth have been varied independently. Glucose-or ammonia-limited cultures were used. The total amount of lipid or relative proportions of neutral lipids and phospholipids did not vary with growth rate or temperature. The relative amounts of the phospholipids varied with growth conditions, but at low temperatures (less than 10° C) the degree of saturation of the fatty acids was strictly controlled and was unaffected by changes in growth rate or nutrient limitation. Under ammonia limitation, increasing the growth rate at a constant temperature had a similar effect on the lipids as decreasing the temperature at a constant growth rate.

2. Protein

Considering the complexity of Gram-negative envelope proteins when examined on SDS/PAGE and their many varied ascribed functions, it is not surprising that the envelope protein composition has been found to vary with growth environment. Much of the work has inevitably been done in batch cultures.[49-56] Of particular interest here are the recent findings on the effect of iron depletion in batch cultures on the outer membrane proteins of Gram-negative bacteria in particular *E. coli*. The contents of three outer membrane proteins have been found to increase due to iron depletion. This has obvious important consequences on the biological properties of the organisms because the proteins concerned have been ascribed properties, not only involving iron uptake, but also phage and colicin binding.[57-62]

Robinson and Tempest[63] studied the effect of nutrient limitation and growth rate on the SDS-PAGE spectra of envelope proteins of *Klebsiella aerogenes*. Distinct patterns of proteins were obtained for each type of nutrient limitation. The effects of carbon and sulfate-limitation were studied in more detail. The amount of protein in the envelopes was growth rate dependent, but envelopes from sulfate-limited organisms always contained less protein (which also had a lower ^{35}S content) than did those from carbon-limited organisms. At low growth rates (D = 0.1 or 0.2 hr^{-1}) the sulphate-limited envelopes contained one major protein (molecular weight 30,000 daltons) whereas the carbon-limited envelopes contained three major proteins (46,000, 38,000, and 28,500 daltons). At higher growth rates (D = 0.6 hr^{-1}) the protein patterns from the two limitations were more similarly. The inner and outer membranes of these organisms were not separated, but on the basis of selective solubility in Triton X-100 both membranes were assumed to have phenotypically variable proteins. The envelope proteins of *Ps. aeruginosa* have similarly been found to vary with growth environment.[64] In contrast to these major changes, Russell et al.[65] found only minor changes in the outer membrane proteins of *Neisseria sicca* due to phosphate-limitation in a chemostat.

Considering the current interest in envelope proteins, especially linking biological properties with the enhancement of certain outer membrane proteins due to iron limitation and the desire to purify individual proteins, it is surprising that the chemostat has not been used more often to accentuate the fluctuations in the contents of envelope proteins observed in batch cultures.

C. Exopolysaccharide

There are a large number of bacteria that secrete extracellular materials either as a capsule or as a slime. These components have been implicated in the pathogenicity of the organism or as a protective measure on the part of the bacterium against various stresses. Much knowledge has been gained about their structure, preparation, and biochemistry of formation. Furthermore they have gained some industrial importance in many fields as food additives and in oil drilling. However little is understood of their physiological role in microorganisms and of course one of the important methods of investigating this area is the use of continuous culture techniques. Of the many important polysaccharides produced by microorganisms only two, alginic acid and xanthan, have been studied in detail using continuous culture methods. The production of these two polymers will now be discussed.

1. Alginic Acid Production

Alginic acid is a 1.4 linked polymer of β-D-mannuronic acid and its 5 epimer, α-L-guluronic acid. The arrangement of the monomers is that of a "block" structure with blocks of one monomer joined to blocks of the other. The polymer is conventionally isolated from brown seaweeds and is a compound of major industrial importance. It is also produced by several bacteria, e.g., strains of *Ps. aeruginosa* and *Azotobacter vinelandii*. The production of alginic acid by this latter organism has been investigated in continuous culture by the Tate and Lyle group.[66] This organism was grown with sucrose as the carbon source and a range of specific respiration rates with phosphate as the limiting substrate. Polysaccharide concentration was essentially constant, decreasing only at very low respiration rates. The production of polysaccharide as a function of dilution rate in this system was also studied. It was found that polysaccharide concentration was highest at low dilution rates, but the rate of polysaccharide production was constant over the range of dilution rates studied.

The effect of differing growth limiting nutrients was also studied. It was found that molybdate and phosphate gave the most favorable specific rates of polysaccharide synthesis. It was also noted that even under sucrose-limited conditions polysaccharide was still produced. This result was most unexpected in that it was thought that the organism would not excrete carbon when it was "carbon" limited.

Pseudomonas NCIB 11264 forms an extracellular slime polysaccharide containing glucose, galactose, mannose, and rhamnose as the monosaccharide units. Acetate and pyruvate residues are also detected in the polysaccharide. Williams and Wimpenny[67] studied the production of this polysaccaride under various conditions. They used single stage continuous culture. The optimum conditions of temperature and pH were found to be 30°C and pH 7.0, respectively. Polysaccharide formation was maximal under nitrogen limitation and dependent on the dilution rate. Phosphate limitation did not enhance the polysaccharide production and no polysaccharide was produced under glucose limitation. The cultures could be run for up to 500 hr without the development of mutant strains, and the culture did not deteriorate in terms of polysaccharide production. The authors concluded that structure of the polysaccharide was constant in all conditions used. However, examination of their data suggests that the pyruvate content of the polymer could vary quite markedly with growth conditions.

2. Xanthan Production

The polymer produced by *Xanthomonas campestris,* Xanthan, has a large number of industrial applications and the growth and polymer production of the organism has been much studied. Xanthan has a cellulose-type backbone structure with side chain residues of mannose, glucuronic acid, and acetate on every second main chain glucose unit. Pyruvate in ketal linkage is attached to the terminal mannose residue of the side chain as follows:

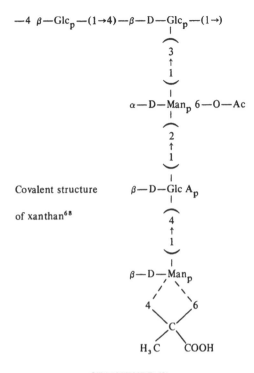

$$-4 \;\beta-Glc_p-(1\to4)-\beta-D-Glc_p-(1\to)$$

Covalent structure

of xanthan[68]

STRUCTURE (9)

This polysaccharide is produced in large quantities by batch fermentation with the growth conditions controlled with respect to pH, oxygen tension, etc.

Silman and Rogovin[69] in their initial studies with single stage continuous fermentation showed that using a complex medium with glucose-excess Xanthan was produced at D = 0.026 hr^{-1} at a rate of 0.36 g/kg/hr. The yield of Xanthan, based on the glucose consumed was 68% and the Xanthan production rate was a function of pH and dilution rate. Further experiments showed however that in longer runs the selection of low producing variants occurred between 6.5 and 8.7 culture generation times. These results with single stage fermenters led this group to suggest the use of multistage fermentation systems to overcome this problem.

These continuous culture studies on *X. campestris* used a complex undefined medium. It would of course be very useful if the continuous culture studies were carried out on a defined medium so that several different limiting nutrients could be studied. It was reported by Starr[70] that *X. campestris* could be grown in a defined medium, but Whistler[71] suggested that a complex medium is required for polysaccharide production. In the studies at MRE on this problem we have found that *X. campestris* and other similar strains, such as *Xanthomonas juglandis,* will grow in a simple salts medium.[72] These organisms will grow continuously in such a medium and under sulfate or phosphate limitation will produce good yields of polysaccharides. There was no culture instability under sulfate limitation and steady-state cultures have been maintained for more than 1000 hr. From these results large scale continuous culture production of the polymer would seem to be feasible. The products from different limiting substrates were essentially similar chemically, but there were marked variations in physical properties. This may be correlated with the variation in the pyruvic acid content of the polymers produced under different conditions.

It was also interesting to note that Xanthan was produced in low concentrations by *X. campestris* when the organism was glucose-limited — a very surprising result if glucose is the substrate used for energy by the cells.

D. Biological Properties of Bacterial Envelopes

The preceding sections illustrate that the envelopes of bacteria are extremely phenotypically variable as a consequence of minor changes in the growth environment. Such changes in envelope structure not surprisingly have pronounced effects on the organisms biological properties — such as pathogenicity, antibiotic sensitivity, phage binding, transport, and attachment. Furthermore growth in the chemostat, used for most of the experiments described in this chapter, more closely mimics the growth of organisms in nature insofar as an "open" system is operating and that the organisms grow at submaximal rates because nutrient limitation may often apply. So chemostat studies would seem to be the system of choice for studying biological properties with respect to disease formation, vaccine production, and ecology.

1. In Relation to Antibiotic Sensitivity

The envelope of a bacterium provides an effective permeability barrier to some antibiotics, thus giving an exclusion mechanism of antibiotic resistance. Our group has extended the work of Brown and Melling[73] who showed that in batch culture magnesium depletion of the growth medium induced resistance of *Ps. aeruginosa* to EDTA and polymyxin B sulfate. Continuous culture of the organisms revealed that the resistance to polymyxin was considerably higher in magnesium limited than carbon- or phosphate limited organisms.[74] Examination of the envelope composition of polymyxin-resistant and sensitive organisms revealed some phenotypic alterations that correlated with polymyxin sensitivity.[64] In particular the phosphate and ethanolamine contents of the envelopes were related to the sensitivity to polymyxin (Table 5). These observations have allowed us to propose a model for resistance of *Ps. aeruginosa* to polymyxin[75] The theory postulates a dual mechanism of antibiotic resistance: (1) Exclusion of polymyxin in the outer membrane by ethanolamine phosphate in the lipopolysaccharide of magnesium-limited organisms thus preventing the antibiotic reaching its target site, the inner membrane. The phosphate-limited organisms have a low ethanolamine phosphate content to bind the positively charged polymyxin, and in carbon-limited organisms the phosphate groups of the ethanolamine phosphate are largely neutralized by magnesium ions and do not bind the polymyxin, (2) Alteration of the target sites of the antibiotics, viz. the phospholipids of the inner membrane.[47] In contrast, Dorrer and Teuber[76] found that phosphate limitation induced resistance to polymyxin in *Ps. fluorescens* due to reduction in the number of target sites of the antibiotics, i.e., replacement of phospholipids in the cytoplasmic membrane with ornithine-containing acidic lipids. The difference between these observations and ours may be due to different organisms used or differences in the degree of phosphate limitation that may effect either or both cytoplasmic and outer membrane.

Finch and Brown have also found that the sensitivities of chemostat cultures of *Ps. aeruginosa* to polymyxin and EDTA, phagocytes, and cationic proteins vary with the nature of the nutrient limitation and the growth rate.[77] Magnesium limitation and growth rate also affected the sensivity of *Ps. aeruginosa* to cold shock.[78] Pickett and Dean[79] found that zinc- and cadmium-tolerant strains of *K. aerogenes* when grown in a chemostat, to abolish the effects of differences in growth rate, were resistant to a number of antibiotics.

Clearly the sensitivity of microorganisms to antimicrobials depends markedly on the growth environment, and much of the variability is a consequence of changes in the envelope permeability barrier.[80] For meaningful in vitro tests on the antibiotic sensitivity of bacteria close attention must be paid to the method of growth if the results are to have any relevance to the in vivo situation.

Table 5

ENVELOPE COMPOSITION OF *PSEUDOMONAS*
AERUGINOSA

Growth limiting nutrient	Content mg/100 mg envelopes			
	Phosphate	KDO*	Ethanolamine	Magnesium
Glucose	2.0	0.50	0.57	0.26
Magnesium	1.8	0.55	0.53	0.11
Phosphate	1.0	0.40	Trace	0.13

Note: The organism was grown in a Porton 0.5 l chemostat at D = 0.1 hr^{-1}.

* 2-Keto-3-deoxy-octonic acid

2. In Relation to Pathogenicity

The involvement of the surface of bacteria in pathogenesis has been reviewed by Smith[81] and only the relevant points will be discussed here. Disease formation by a microbe is dependent upon determinants of pathogenicity, many of which are surface properties and effect (1) entry to the host, (2) multiplication in the host, (3) resistance to host defenses, and (4) damage to the host. Obviously alteration in the surface of microorganisms can alter their pathogenicty. It seems that multiplication of bacteria in animal tissues occurs at a submaximal, substrate-limited rate so that the chemostat would appear to be an important tool for studying pathogenicity. Furthermore changes in the surface of microorganisms seen in the chemostat may also occur in vivo during the disease process.

The variation in sensitivity of organisms to lysozyme[4,16,30,31] and phagocytes and cationic proteins[77] has already mentioned. Pearson and Ellwood[82] found that the toxicity of envelopes of *K. aerogenes* varied with the growth environment in the chemostat—envelopes from a glycerol-limited culture grown at high dilution rates being most toxic, those from magnesium-limited cultures being least toxic. Presumably the structure of the lipopolysaccharide in the envelopes is varying, but no correlation could be found between toxicity and KDO or 3-hydroxymyristic acid contents. Similarly the toxicity of whole cells of *E. coli* showed growth environment-dependent variability.[83] For the same dilution rate glycerol-limited organisms were the most toxic and sulfate-limited organisms the least. Again envelope structure differences were assumed to account for the results.

Another bacterium of interest in our laboratory is *Bordetella pertussis*, the causative organism of whooping cough. Lacey [84] described antigenic modulation of *B. pertussis* dependent upon the ionic composition of the growth medium. The 'X' (for xanthic) mode elicits an array of unique immunological and physiopathological responses in man and certain laboratory animals. The substances responsible for these effects (including the protective antigen, histamine sensitizing factor, lymphocytosis promoting factor, hemagglutinin, and up to eight heat labile agglutinogens) appear to be envelope associated. The 'C' (for cyanic) mode, which is a phenotypic variant produced when, for example, the sodium ions in the growth medium are replaced by magnesium ions, are deficient in these biological properties. Parton and Wardlaw[50] found that the envelopes of 'C' mode organisms when examined by SDS-PAGE were deficient in two major (and a number of minor) envelope proteins with molecular weights of 30,000 and 33,500 daltons. Variation in the nicotinic acid in the medium can effect similar changes.[85] We have also found a strong correlation between the mouse protective activity of *B. pertussis* and the presence of these two envelope proteins.[86] Indeed whether a preparation was cells, envelopes, or an envelope extract, the preparation was only protective when the two (possible three) envelope proteins were present. These variable envelope proteins have been tentatively assigned to the outer envelope membrane on chemical considerations. Continuous cultivation of the organism in chemostats produced the 'X' to 'C' phenotypic variation when the medium was changed from a complex to a chemical defined medium or on prolonged growth.[86-88]

Iron is essential for bacterial growth and in vivo a reduction in available iron (due to uptake by host transferrin and lactoferrin) has a pronounced effect on bacterial pathogenicity. The pathogen must be able to compete with the host for available iron. The effect of iron depletion on envelope structure of bacteria has already been discussed in Section IIIB2. Iron limitation is chemostats would seem to be the obvious approach for studying the effects of iron depletion on bacterial pathogenicity.

Considering the obvious advantages of controlled growth in the chemostat, it is surprising that vaccines are not produced by continuous culture to obtain cells of highest possible potency and minimum toxicity.

IV. INTRACELLULAR MACROMOLECULES

Space does not allow a detailed account of the effect of continuous culture on intracellular molecules. We shall thus limit ourselves to brief discussions on the overall gross macromolecular compositions in the hope that more specific changes will be dealt with in other chapters.

A. Energy Reserve Polymers

Energy reserve polymers will only be dealt with briefly here, as an excellent review by Dawes and Senior[89] examines the subject in depth. Three criteria have to be established before a compound can have an energy storage function assigned to it. (1) The compound must be accumulated under conditions of excess supply of energy from exogenous sources, (2) the compound must be utilized when the energy supply from exogenous sources is insufficient for optimal maintenance of the cell, and (3) the compound is degraded to produce utilizable energy for the cell conferring an advantage on those cells containing the polymer over those that do not.

Polyphosphate (occurring in "volutin" granules with RNA, lipid, protein, and magnesium ions) will not be discussed here except to mention that the synthesis of polyphosphate is obviously related to nucleic acid synthesis and the sulfur limitation may be important in its formation.[89]

Polyphosphate (PHB) is an excellent storage compound, as it exists in a highly reduced state as a virtually insoluble crystalline polymer exerting neglible osmotic pressure. The structure is as follows:

$$HO-\underset{\underset{CH_3}{|}}{CH}-CH_2-\underset{\underset{O}{\|}}{C}\left[-O-\underset{\underset{CH_3}{|}}{CH}-CH_2-\underset{\underset{O}{\|}}{C}\right]_n-O-\underset{\underset{CH_3}{|}}{CH}-CH_2-COOH$$

STRUCTURE (10)

n is usually 600 to 2500 and PHB can account for almost 90% of the dry weight of a cell. Much of the work on PHB has been reviewed by Dawes and Senior[89] who cite most of the relevant references that will not be recited here. The formation of PHB appears to be related to both nutrient limitation and growth rate. With *Azotobacter beijerinckii* oxygen limitation is the most important factor for the acumulation of PHB by cells. Under these conditions a build up of $NAD(P)H_2$ leads to a slower utilization of acetyl-CoA by the tricarboxylic cycle, and acetyl-CoA is channelled into PHB formation. These conditions inhibit the the utilization of PHB. Synthesis of PHB is restricted when high intracellular concentrations of NAD(P) occur (such as in the relaxation of oxygen limitation). Breakdown occurs when the steady-state concentration of acetyl-CoA decreases and that of coenzyme-A increases due to the supply of glucose catabolites being restricted. Similar considerations apply to the regulation of PHB synthesis in organisms that accumulate the polymer under conditions of growth limitation other than oxygen, (e.g., *Bacillus megaterium* with nitrogen, sulfur and potassium limitation and *Hydrogenomonas* under nitrogen limitation). PHB synthesis does not occur under nitrogen limitation in *A. beijerinckii* in contrast to many organisms—possibly due to respiratory protection of the nitrogenase system in this organism. This problem has been recently examined by gradually increasing the oxygen inflow in oxygen-limited cultures.[90] With ammonia-grown cultures as oxygen inflow is increased, there is a fall in PHB content, then a new maximum occurs, and finally a decrease to negligible values as ammonia limitation is established. With nitrogen-grown cultures there is a steady decrease in PHB content with increasing oxygen concentration.

Glycogen, which is an α $(1{\rightarrow}4)$ glucan with occasional α $(1{\rightarrow}6)$ branch linkages was known to be formed intracellularly when organisms were grown in a carbohydrate-rich medium in batch culture, and is considered to be an energy reserve polymer.

The formation of glycogen in bacteria growing in a chemostat was first studied by Holme.[91] He showed that *E. coli* B grown nitrogen limited with glucose as the carbohydrate source formed the maximum amount of glycogen at low growth rates. Indeed there was an inverse relationship between growth rate and glycogen formation. Holme and co-workers showed that the glycogen formed consisted of a high and low molecular weight fraction. Interestingly the slow growing organism contained the largest amount of the high molecular weight material, whereas the faster grown organism contained less glycogen, but this was predominantly of the low molecular weight type. Bacteria use ADP-glucose as the glucose donor in the synthesis of glycogen. The relationship of ADP to ATP and AMP is defined as the energy charge of the cell. When this is high, the cell contains high amounts of ATP and it seems likely in this situation that glycogen will be synthesized as an energy sink, whereas when the energy charge is low any glycogen in the cell will be degraded to produce more ATP.

Many organisms form glycogen from glucose when grown in a glucose-rich medium, but usually this only occurs in the later stages of growth in batch culture. In *Streptococcus mitis* and other oral streptococci, glycogen synthesis occurs in all stages of growth including the exponential phase. Furthermore nitrogen- limited populations in phosphate buffer with added glucose accumulated large quantities of polysaccharide. *S. mutans* FA1 produced up to 50% of its dry weight as "glycogen" when grown in glucose excess in batch culture.[92]

It has been suggested that this accumulation of glycogen could act as a virulence factor in causing dental caries, as acid would be produced for long periods by degradation of the accumulated glycogen. Furthermore bacteria in dental plaque contain granules of carbohydrate. But recently it has been shown that cariogenic strains of streptococci do not accumulate large amounts of glycogen when grown in a glucose excess medium.[93]

B. Protein and Nucleic Acid

The first comprehensive study of environmental effects in continuous culture on the cellular contents of protein and nucleic acid was made by Herbert[3] in 1961. From this fundamental study it became apparent "that it is virtually meaningless to speak of the chemical composition of a microorganism without at the same time specifying the environmental conditions that produced it''. For example the RNA content of an organism can vary from 1.5 to 40%. As nucleic acid and protein synthesis are closely associated, they will be discussed together. The DNA and protein contents of the biomass (*K. aerogenes* and *B. megaterium*) decreased slightly with increasing growth rate, whereas the RNA content and mean cell mass increased markedly (see Figure 3). The same striking pattern of change was found to occur, whatever the nature of the nutrient that was limiting.

The chemostat has been proved to be a useful tool for the study of the synthesis of RNA and protein as indicated by the following experiments. Nitrogen limitation obviously has an effect on protein synthesis and phosphate limitation on nucleic acid synthesis.[94] Magnesium limitation has a marked effect on the RNA composition of bacteria. Sykes and Tempest[95] found the magnesium-limited *Pseudomonas* strain C-1B contained more RNA than did carbon-limited organisms. The additional RNA was not ribosomal as the organisms had equal ribosome contents. This study was extended further by Sykes and Young[96,97] in an examination of the effects of carbon and magnesium limitation in continuous culture on the ribosomes and RNA of *K. aerogenes*. For both limitations the ratios of 50S/30S ribosomes and 23S/16S RNA and the sedimentation constants of these components were equal and invariant with growth rate. Under carbon limitation, ribosomal RNA accounted for 89% of the total hot-acid soluble RNA at high growth rates and 68% at low growth rates; 4S (transfer) RNA accounted for the remainder. At all growth rates ribosomal RNA accounted for 89% of the total hot-acid soluble RNA of magnesium-limited organisms. The efficiency of ribosomal RNA in protein synthesis (measured by $\frac{\text{protein D}}{\text{RNA}}$) was found to be not growth rate independent, increasing under carbon-limitation as D was increased up to a value of 0.5 hr^{-1}, and thereafter remaining constant—the further increase in growth rate being dependent on increased ribosome production. In constrast the magnesium-limited system showed a constantly rising efficiency of ribosomal RNA in protein synthesis with increasing growth rate. It is interesting to correlate these results with the observation in batch cultures that the protein composition of ribosomes is affected by growth rate.[98,99] The rate of ribosomal RNA synthesis in bacteria is correlated with protein synthesis and a function of the rel A gene. Atherly[100] examined the regulation of RNA synthesis in cultures of rel A$^-$ and rel A$^+$ (relaxed and stringent RNA synthesis control respectively) strains of *E. coli* in a chemostat with growth limited by an essential amino acid. The results suggested that when growth is limited by an initiator of protein synthesis, i.e., methionine, the RNA regulation occurs in stringent and relaxed strains. When other amino acids were limiting, errors occurred during translation in the relaxed strains resulting in misread proteins.

An interesting observation has been made on the stability of plasmids in *E. coli* growing in a chemostat. Melling et al.[101] found that the R-factor RP1 was stable in pure cultures of R-factor bearing *E. coli* under glucose, magnesium, or phosphate limitation at dilution rates varying between 0.05 and 1.0 hr^{-1}. However in mixed culture studies a very small number of the R-factor-free parent organisms rapidly outgrew and replaced the R-factor bearing organism under phosphate limitation, but not under carbon or magnesium limitation. The rate of loss of R-factor bearing organisms under phosphate limitation was almost equivalent to the rate of washout of a nongrowing population. Interestingly a small percentage of R-factor bearing organisms remained in the culture vessel even after prolonged periods of growth. Clearly the presence of the plasmid is a severe handicap to the organism under phosphate limitation. Presumably the R-factor bearing organism is less efficient than the parent at phosphate

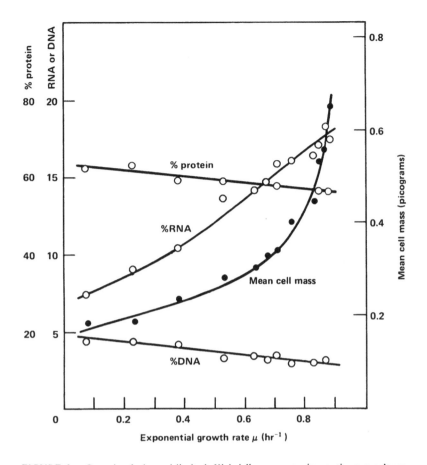

FIGURE 3. Growth of glycerol-limited *Klebsiella aerogenes* in continuous culture; protein and nucleic acid contents and mean cell mass as a function of growth rate. Nucleic acid and protein contents are expressed as percentage of cell dry weight. Mean cell mass = dry weight per milliliter divided by total cell count per milliliter. (From Herbert, D., *Symp. Soc. Gen. Microbiol.*, XI, Cambridge University Press, London, 391, 1961. With permission.)

uptake. The results suggest a whole new field for the study of plasmid stability and replication, with relevance to the prevalence of R-factors in nature and the safety of genetic engineering experiments.

V. SUMMARY AND CONCLUSIONS

This chapter set out to review the work using continuous culture techniques in studies on macromolecular composition of microorganism with particular reference to bacterial cell envelopes. The initial studies in this area showed quite clearly that independent of the nature of the limiting nutrient the total RNA of a cell increased with increasing growth rate, but the total DNA remained roughly constant. This makes sense in that the DNA of the cell, being the genetic information, remained invariant, but the RNA had to increase to meet the demands for increased biosynthesis.

Investigation of different nutrient limitations and growth rates on the cationic compositions of organisms showed that a relationship existed between the RNA and magnesium contents of cells. This was true for representative types of Gram-positive and Gram-negative bacteria. However excess phosphorus observed in Gram-positive bacteria when grown under magnesium limitation was lost when the organisms were grown phosphate limited. This observation led to the discovery of the variability of Gram-positive bacterial cell walls and using chemostat techniques it was possible to show that the wall behaved as a functional organelle of the cell. This functional role also brings into focus the concept of the wall behaving in a dynamic way in response to environmental changes.

The ideas of cell wall turnover in Gram-positive bacteria seem to be clearly demonstrable, but in Gram-negative bacteria the evidence is not so clear. It is a common observation that bacteria grown in batch culture secrete lipopolysaccharide into the culture fluid and this at least is consistent with the idea of the dynamic nature of the Gram-negative bacterial cell surface.

Cell wall turnover may play an important role in the surveillance of available food sources by organisms in the natural environment. For example environmental changes leading to wall turnover may allow the release of degradative enzymes to scavenge for nutrients in the new environment. Growth rates of bacteria in natural environments have not been well characterized, but there is evidence that growth occurs at rates considerably slower than those obtained by laboratory batch culture methods. This suggests that natural environments are not ideal for bacterial growth. Hence studies of bacteria growing slowly are essential to the understanding of bacteria in the natural environment and these can most conveniently be carried out in the laboratory using continuous culture techniques. Bacteria cultivated at slow growth rates may, for example, prove to be more potent vaccines than conventionally batch grown organisms because antigenically they more closely resemble the pathogen in its natural environment.

REFERENCES

1. **Evans, C. G. T., Herbert, D., and Tempest, D. W.,** The continuous cultivation of microorganisms. II. The construction of a chemostat, in *Methods in Microbiology,* Vol. 2, Norris, J. R. and Ribbons, D. W., Eds., Academic Press, New York, 1970, 277.
2. **Tempest, D. W.,** Quantitative relationships between inorganic cations and anionic polymers in growing bacteria, *Symp. Soc. Gen. Microbiol.,* XIX, 87, 1969.
3. **Herbert, D.,** The chemical composition of microorganisms as a function of their environment, *Symp. Soc. Gen. Microbiol.,* XI, 391, 1961.
4. **Ellwood, D. C. and Tempest, D. W.,** Effect on environment on bacterial wall content and composition, *Adv. Microb. Physiol.,* 7, 83, 1972.
5. **Mulder, E. G. and Antheuisse, J.,** Morphologie, physiologie et ecologie des Arthrobacter, *Ann. Inst. Pasteur, Paris,* 105, 46, 1963.
6. **Luscombe, B. M. and Gray, T. R. G.,** Effect of varying growth rate on the morphology of Arthrobacter, *J. Gen. Microbiol.,* 69, 433, 1971.
7. **Luscombe, B. M. and Gray, T. R. G.,** Characteristics of Arthrobacter grown in continuous culture, *J. Gen. Microbiol.,* 82, 213, 1974.
8. **Ellwood, D. C., Hunter, J. R., and Longyear, V. M. C.,** Growth of *Streptococcus mutans* in a chemostat, *Arch. Oral Biol.,* 19, 659, 1974.
9. **Armstrong, J. J., Baddiley, J., Buchanan, J. G., Carss, B., and Greenburg, G. R.,** Isolation and structure of ribitol phosphate derivatives (teichoic acids) from bacterial cell walls, *J. Chem. Soc.,* p. 4344, 1958.
10. **Archibald, A. E.,** The structure, biosynthesis and function of teichoic acids, *Adv. Microb. Physiol.,* 11, 53, 1974.

11. **Branefors-Helander, Paula, Erbing, Christina, Kenne, L., and Lingberg, B.,** The structure of the capsular antigen from *Haemophilus Influenzae*Type A, *Carbohydr. Res.,*56, 117, 1977.
12. **Orskov, Ida, Orskov, F., Jann, Barbara, and Jann, K.,** Serology, chemistry and genetics of O and K antigens of *Escherichia coli*(see p. 679), *Bacteriol. Rev.,*41, 667, 1977.
13. **Ellwood, D. C.,** The wall content and composition of *Bacillus subtilis*var *niger*grown in a chemostat, *Biochem. J.,*118, 367, 1970.
14. **Baddiley, J., Hancock, I., and Sherwood, P. M. A.,** X-ray photoelectron studies of magnesium ions bound to the cell walls of Gram-positive bacteria, *Nature (London),*243, 43, 1973.
15. **Anderson, A. J., Green, R. S., and Archibald, A. R.,** Specific determination of ribitol teichoic acid in whole bacteria and isolated walls of *Bacillus subtilis*W23,*Carbohydr. Res.,*57, C7, 1977.
16. **Archibald, A. R. and Heckels, J. E.,** Alterations in the composition and bacteriophage-binding properties of walls of *Staphylococcus aureus* H grown in simplified defined media, *Biochim. Biophys. Acta,* 406, 60, 1975.
17. **Forseberg, C. W., Wyrich, P. B., Ward, J. B., and Rogers, H. J.,** Effect of phosphate-limitation on the morphology and wall composition of *Bacillus licheniformis* and its phosphoglucomutase-deficient mutants, *J. Bacteriol.,*113, 969, 1973.
18. **Heckels, J. E., Archibald, A. R., Baddiley, J.,** Studies on the linkage between teichoic acid and peptidoglycan in a bacteriophage-resistant mutant of *Staphylococcus aureus* H, *Biochem. J.,* 149, 637, 1975.
19. **Wicken, A. J. and Knox, K. W.,** Lipoteichoic acids — a new class of bacterial antigens, *Science,* 187, 1161, 1975.
20. **Wicken, A. J. and Knox, K. W.,** Biological properties of lipoteichoic acids, *Microbiology,* A.S.M. Publications, 1977, 360.
21. **Mauk, J., Chan, L., and Glaser, L.,** Turnover of the cell wall of Gram-positive bacteria, *J. Biol. Chem.,* 246, 1820, 1971.
22. **Archibald, A. R. and Coapes, H. E.,** Bacteriophage SP50 as a marker for cell wall growth in *Bacillus subtilis, J. Bacteriol.,* 125, 1195, 1976.
23. **Young, F. E.,** Variation in the chemical composition of the cell walls of *Bacillus subtilis* during growth in different media, *Nature (London),* 207, 104, 1965.
24. **Chin, Theresa, Burger, M. M., and Glaser, L.,** Synthesis of teichoic acids VI. The formation of multiple wall polymers in *Bacillus subtilis*W23, *Arch. Biochem. Biophys.,* 116, 358, 1966.
25. **Wright, J. and Heckels, J. E.,** The teichuronic acid of cell walls of *Bacillus subtilis* W23 grown in a chemostat under phosphate-limitation, *Biochem. J.,* 147, 187, 1975.
26. **Archibald, A. R.,** personal communication.
27. **Henning, U.,** Determination of cell shape in bacteria, *Annu. Rev. Microbiol.,*29, 45, 1975.
28. **Schliefer, K. H. and Kandler, O.,** Peptidoglycan types of bacterial cell walls and their taxonomic implications, *Bacteriol. Rev.,*36, 407, 1972.
29. **Schliefer, K. H., Hammes, W. P., and Kandler, O.,** Effect of endogenous and exogenous factors on the primary structures of bacterial peptidoglycan, *Adv. Microb. Physiol,* 13, 245, 1975.
30. **Ellwood, D. C.,** The effect of growth conditions on the mucopeptide composition of walls of *Bacillus subtilis*W23, *Biochem. J.,*127, 73P, 1972.
31. **Johnson, K. G. and Campbell, J. N.,** Effect of growth conditions on peptidoglycan structure and susceptibility to lytic enzymes in cell walls of *Micrococcus sodonensis, Biochemistry,* 11, 277, 1972.
32. **De Rosa, M., Gambacorta, A., and Bu'lock, J. D.,** Effects of pH and temperatures on the fatty acid composition of *Bacillus acidocaldarius, J. Bacteriol.,* 117, 212, 1974.
33. **Van Schaik, F. W. and Veerkamp, J. H.,** Biochemical changes in *Bifidobacterium bifidum* var *pennsylvanicus* after cell wall inhibition. VIII. Composition and metabolism of phospholipids at different stages and conditions of growth, *Biochim. Biophys. Acta,* 388, 213, 1975.
34. **Minnikin, D. E. and Abdolrahimzadeh, H.,** Effect of pH on the proportions of polar lipids in chemostat cultures of *Bacillus subtilis, J. Bacteriol.,* 120, 999, 1974.
35. **Drucker, D. B., Griffith, C. J., and Melville, T. H.,** Fatty acid fingerprints of *Streptococcus mutans* grown in a chemostat, *Microbios,* 7, 17, 1973.
36. **Drucker, D. B., Griffith, C. J., and Melville, T. H.,** Fatty acid fingerprints of streptococci:variability due to carbohydrate source, *Microbios,* 9, 187, 1974.
37. **Drucker, D. B., Griffith, C. J., and Melville, T. H.,** The influence of vitamin and magnesium limitations on fatty acid fingerprints of chemostat grown *Streptococcus* Sp SS., *Microbios,* 10, 13, 1974.
38. **Drucker, D. B., Griffith, C. J., and Melville, T. H.,** Fatty fingerprints of some chemostat-grown streptococci with computerized data analysis, *Microbios Letters,* 1, 31, 1976.
39. **Schnaitman, C. A.,** Effect of ethylenediamine tetraacetic acid, Triton X-100, and lysozyme on the morphology and chemical composition of isolated cell walls of *Escherichia coli, J. Bacteriol,* 108, 553, 1971.

40. **Osborn, M. J., Gander, J. E., Parisi, E., and Carson, J.,** Mechanism of assembly of the outer membrane of *Salmonella typhimurium.* Isolation and characterization of cytoplasmic and outer membranes, *J. Biol. Chem.,* 247, 3962, 1972.

41. **Hindennach, I. and Henning, U.,** The major proteins of the *Escherichia coli* outer cell envelope membrane. Preparative isolation of all major membrane proteins, *Eur. J. Biochem.,* 59, 207, 1975.

42. **Braun, V.,** Covalent lipoprotein from the outer membrane of *Escherichia coli, Biochim. Biophys. Acta,* 415, 335, 1975.

43. **Lugtenberg, B., Bronstein, H., van Selm, N., and Peters, R.,** Peptidoglycan-associated outer membrane proteins in Gram-negative bacteria, *Biochim. Biophys. Acta,* 465, 571, 1977.

44. **Henning, U. and Haller, I.,** Mutants of *Escherichia coli* K12 lacking all 'major' proteins of the outer cell envelope membrane, *FEBS Lett.,* 55, 161, 1975.

45. **Minnikin, D. E. and Abdolrahimzadeh, H.,** The replacement of phosphatidyl-ethanolamine and acid phospholipids by an ornithine-amide lipid and a minor phosporus-free lipid in *Pseudomonas fluorescens* N.C.M.B. 129 *FEBS Lett.,* 43, 257, 1974.

46. **Minnikin, D. E. Abdolrahimzadeh, H., and Baddiley, J.,** Replacement of acidic phospholipids by acidic glycolipids in *Pseudomonas diminuta, Nature (London),* 249, 268, 1974.

47. **Minnikin, D. E.,** personal communication, 1974.

48. **Gill, G. O.,** Effect of growth temperature on the lipids of *Pseudomonas fluorescens, J. Gen. Microbiol.,* 89, 293, 1975.

49. **Starka, V.,** Cell envelope protein of dividing and non-dividing cells of *Escherichia coli, FEBS Lett.,* 16, 223, 1971.

50. **Parton, R. and Wardlaw, A. C.,** Cell-envelope proteins of *Bordetella pertussis, J. Med. Microbiol.,* 8, 47, 1975.

51. **Ames, G. F-L.,** Resolution of bacterial proteins by polyacrylamide gel electrophoresis on slabs. Membrane, soluble, and periplasmic fractions, *J. Biol. Chem.,* 249, 634, 1974.

52. **Schnaitman, C. A.,** Outer membrane proteins of *Escherichi coli* IV. Differences in outer membrane proteins due to strain and cultural differences, *J. Bacteriol.,* 118, 454, 1974.

53. **Gilleland, H. E., Stinnett, J. D., and Eagon, R. G.,** Ultrastructural and chemical alteration of the cell envelope of *Pseudomonas aeruginosa,* associated wtih resistance to ethylenediamine tetraacetate resulting from growth in Mg^{2+} - deficient medium, *J. Bacteriol.,* 117, 302, 1974.

54. **Spencer, M. E. and Guest, J. R.,** Proteins of the inner membrane of *Escherichia coli* changes in composition associated with anaerobic growth and fumarate reductase amber mutation, *J. Bacteriol.,* 117, 954, 1974.

55. **van Alphen, W. and Lugtenberg, B.,** Influence of osmolarity of the growth medium on the outer membrane protein pattern of *Escherichia coli, J. Bacteriol.,* 131, 623, 1977.

56. **Lugtenberg, B., Peters, R., Bernheimer, H., and Berendsen, W.,** Influence of cultural conditions and mutations on the composition of the outer membrane proteins of *Escherichia coli, Mol. Gen. Genet.,* 147, 251, 1976.

57. **Braun, V., Hancock, R. E. W., Hantke, K., and Hartman, A.,** Functional organisation of outer membrane of *Escherichia coli* phage and colicin receptors as components of iron uptake systems, *J. Supramol. Struct.,* 5, 37, 1976.

58. **Ichihara, S. and Mizushima, S.,** Involvement of outer membrane proteins in enterochelin-mediated iron uptake in *Escherichia coli, J. Biochem.,* 81, 749, 1977.

59. **Benner, R. L. and Rothfield, L. I.,** Genetic and physiological regulation of intrinsic proteins of the outer membrane of *Salmonella typhimurium, J. Bacteriol.,* 127, 498, 1976.

60. **Pugsley, A. P. and Reeves, P.,** Increased production of the outer membrane receptors for colicins B, D, and M by *Escherichia coli* under iron starvation, *Biochem. Biophys. Res. Commun.,* 70, 846, 1976.

61. **Datta, B. D., Arden, B., and Henning, U.,** Major proteins of the *Escherichia coli* outer cell envelope membrane as bacteriophage receptors, *J. Bacteriol.,* 131, 821, 1977.

62. **Soucek, S. and Konisky, J.,** Normal iron-enterochelin uptake in mutants lacking colicin I. Outer membrane receptor protein of *Escherichia coli, J. Bacteriol.,* 130, 1399, 1977.

63. **Robinson, A. and Tempest, D. W.,** Phenotypic variability of the envelope proteins of *Klebsiella aerogenes, J. Gen. Microbiol.,* 78, 361, 1973.

64. **Robinson, A., Melling, J., and Ellwood, D. C.,** Effect of growth environment on the envelope composition of *Pseudomonas aeruginosa, Proc. Soc. Gen. Microbiol.,* 1, 61, 1974.

65. **Russell, R. R. B., Johnson, K. G., and McDonald, J.,** Envelope proteins in *Neisseria, Can. J. Microbiol.,* 21, 1519, 1975.

66. **Deavin, L., Jarman, T. R., Lawson, C. J., Righelato, R. C., and Slocombe, S.,** The production of alginic acid by *Azobacter vinelandii* in batch and continous culture, in *Extracellular Microbial Polysaccharides,* Sandford, P. A. and Laskin, A., Eds., American Chemical Symposium No. 45, 1977, 14.

67. **Williams, A. G. and Wimpenny, J. W. T.**, Exopolysaccharide production by *Pseudomonas* NCIB11264 grown in continuous culture, *J. Gen. Microbiol.*, 104, 47, 1978.
68. **Jansson, P. E., Kenne, L., and Linberg, B.**, Structure of the extracellular polysaccharide from *Xanthomonus campestris, Carbohydr. Res.*, 45, 275, 1975.
69. **Silman, R. W. and Rogovin, P.**, Continuous fermentation to produce Xanthan biopolymer:effect of dilution rate, *Biotechnol. Bioeng.*, XIV, 23, 1972.
70. **Starr, M. P.**, The nutrition of phytopathogenic bacteria, *J. Bacteriol.*, 51, 131, 1946.
71. **Whistler, R. L.**, in *"Industrial Gums"*, Academic Press, New York, 1973.
72. **Evans, C. S. T., Ellwood, D. C., and Yeo, R. G.**, *Br. Pat.*, 46784/75.
73. **Brown, M. R. W. and Melling, J.**, Role of divalent cations in the action of polymyxin B and EDTA on *Pseudomonas aeruginosa, J. Gen. Microbiol.*, 59, 263, 1969.
74. **Melling, J., Robinson, A., and Ellwood, D. C.**, Effect of growth environment in a chemostat on the sensitivity of *Pseudomonas aeruginosa* to polymyxin B sulphate, *Proc. Soc. Gen. Microbiol.*, 1, 61, 1974.
75. **Melling, J., Ellwood, D. C., and Robinson, A.**, unpublished observation.
76. **Dorrer, E. and Teuber, M.**, Induction of polymyxin resistance in *Pseudomonas fluorescens* by phosphate-limitation, *Arch. Microbiol.*, 114, 87, 1977.
77. **Finch, J. E. and Brown, M. R. W.**, The effect of growth environment on the killing of chemostat grown *Pseudomonas aeruginosa* by phagocytes and cationic proteins, *Proc. Soc. Gen. Microbiol.*, 3, 82, 1976.
78. **Kenward, M. A., Brown, M. R. W., and Fryer, J. J.**, The effect of carbon- or magnesium-limitation upon the survival and drug resistance of cold-shocked *Pseudomonas aeruginosa, Proc. Soc. Gen. Microbiol.*, 3, 83, 1976.
79. **Pickett, A. W. and Dean, A. C. R.**, Antibiotic resistance of cadmium- and zinc-tolerant strains of *Klebsiella (Aerobacter) aerogenes* growing in glucose-limited chemostats, *Microbios Letters*, 1, 165, 1976.
80. **Brown, M. R. W.**, The role of the cell envelope in resistance, in *Resistance of Pseudomonas aeruginosa*, Brown, M. R. W., Ed., Wiley, 1975, chap. 3, 71.
81. **Smith, H.**, Microbial surfaces in relation to pathogenicity, *Bacteriol. Rev.*, 41, 1977, 475.
82. **Pearson, A. D. and Ellwood, D. C.**, The effect of growth conditions on the chemical composition and endotoxicity of walls of *Aerobacter aerogenes*, NCTC 418, *Biochem. J.*, 127, 72, 1972.
83. **Pearson, A. D. and Ellwood, D. C.**, Growth environment and bacterial toxicity, *J. Med. Microbiol.*, 7, 391, 1974.
84. **Lacey, B. W.**, Antigenic modulation of *Bordetella pertussis, J. Hyg.*, 58, 57, 1960.
85. **Wardlaw, A. C., Parton, R., and Hooker, M. J.**, Loss of protective antigen, histamine sensitizing factor, and envelope polypeptides in cultural variants of *Bordetella pertussis, J. Med. Microbiol.*, 9, 89, 1976.
86. **Robinson, A., Manchee, R. J., and Ellwood, D., C.**, Mouse protection potency of batch and continously cultured *Bordetella pertussis, Proc. Soc. Gen. Microbiol.*, 4, 9, 1976.
87. **Manchee, R. J., Robinson, A., and Ellwood, D. C.**, Amino acid utilization patterns of *Bordetella pertussis* in continuous culture, *Proc. Soc. Gen. Microbiol.*, 4, 9, 1976.
88. **van Hemert, P.**, Vaccine production as a unit process, *Prog. Ind. Microbiol.*, 13, 151, 1974.
89. **Dawes, E. A. and Senior, P. J.**, The role and regulation of energy reserve polymers in micro-organisms, *Adv. Microb. Physiol.*, 10, 135, 1973.
90. **Ward, A. C., Rowley, B. I., and Dawes, E. A.**, Effect of oxygen- and nitrogen-limitation on poly-β-hydroxybutyrate biosynthesis in ammonia-grown *Azotobacter beijerinckii, J. Gen. Microbiol.*, 102, 61, 1977.
91. **Holme, T.**, Continuous culture studies on glycogen synthesis in *Escherichia coli* B, *Acta Chem. Scanda.*, 11, 762, 1957.
92. **Mattingly, S. J., Dipensio, J. R., Higgens, M. L., and Shockman, G. D.**, Unbalanced growth and macromolecular synthesis in *Streptococcus mutans* FA-1, *Infect. Immun.*, 13, 941, 1976.
93. **Hamilton, I. R.**, Intracellular polysaccharide synthesis by cariogenic microoganisms, *Microbial Aspects of Dental Caries*, Stiles, H. M., Loesche, W. J., and O'Brien, T. C., Eds., Sp. Supp. Microbiology Abstracts Vol. 1, 683, 1976.
94. **Cooney, C. L. and Wang, D. I. C.**, Transient response of *Enterobacter aerogenes* under a dual nutrient limitation in a chemostat, *Biotechnol. Bioeng.*, 18, 189, 1976.
95. **Sykes, J., and Tempest, D. W.**, The effect of magnesium- and carbon-limitation on the macromolecular organisation and metabolic activity of *Pseudomonas* strain C-1B, *Biochim. Biophys. Acta*, 103, 93, 1965.
96. **Sykes, J. and Young, T. W.**, Studies on the ribosomes and ribonucleic acids of *Aerobacter aerogenes* grown at different rates in carbon-limited continuous culture, *Biochim. Biophys. Acta*, 169, 103, 1968.

97. **Young, T. W. and Sykes, J.,** Studies on the ribosomes and ribonucleic acids of *Aerobacter aerogenes* grown at different rates in magnesium-limited continuous culture, *Biochim. Biophys. Acta,* 169, 117, 1968.
98. **Deusser, E. and Wittman, H. G.,** Ribosomal proteins: variation in the protein composition of *Escherichia coli* ribosomes as a function of growth rate, *Nature (London),* 238, 269, 1972.
99. **Milne, A. N., Mak, W. W-N., and Wong, J. T-F.,** Variation of ribosomal proteins with bacterial growth rate, *J. Bacteriol.,* 122, 89, 1975.
100. **Atherly, A. G.,** Ribonucleic acid regulation in amino acid-limited cultures of *Escherichia coli* grown in a chemostat, *J. Bacteriol.,* 120, 1322, 1974.
101. **Melling, J., Ellwood, D. C., and Robinson, A.,** Survival of R-factor carrying *Escherichia coli* in mixed cultures in the chemostat, *FEMS Microbiol. Lett.,* 2, 87, 1977.

Chapter 3

REGULATION OF ENZYME SYNTHESIS AS STUDIED IN CONINUOUS CULTURE*

Abdul Matin

TABLE OF CONTENTS

---

* This chapter was submitted in November 1978.

I. INTRODUCTION

The continuous culture technique (chemostat) is vastly superior to batch cultures for studies on microbial regulatory mechanisms. In batch cultures, the environmental conditions change continuously and uncontrollably during growth, leading in turn to a continuous alteration of the culture phenotype, and complicating interpretation of results. Further, it is often not possible to alter the composition of the growth environment without concurrent alteration of the culture growth rate, and since the latter can independently influence an organisms' physiology, the relationship between the environment and the culture physiology can at best be tenuously established. In contrast, in the chemostat growth occurs under steady-state conditions in which the growth environment and consequently the culture phototype become fixed and time independent, and the use of this technique permits alteration of growth environment independent of growth rate. In addition, growth in the chemostat occurs under nutrient limitation and since this is likely to be the mode of microbial growth in most natural environments,[1] studies employing this technique probably have a greater ecological relevance than those relying on batch cultures.

In this chapter, I will discuss those continuous culture studies that provide insights into the molecular mechanism of enzyme synthesis, as well as those dealing with the special regulatory phenomena that enable microorganisms to adapt to nutrient-limited environments.

II. INFLUENCE OF DILUTION RATE ON ENZYME SYNTHESIS

A. Bacteria

Table 1 summarizes the influence of culture dilution rate (D) on bacterial enzyme activity during carbon-limited growth. Most of these activities were measured in extracts of cells grown at different D values, and it was assumed that the observed differences reflected variation in the amount of enzyme synthesized rather than modulation of the activity of existing enzymes. In cases where proper controls were run,[3] the results were consistent with this interpretation. It would be of considerable interest, however, to correlate the enzyme levels with the levels of messenger RNA in the cells at differerent D values. In most of these studies, no information was provided on whether or not oscillations occurred in the enzyme levels in steady-state cultures. Such oscillations have been reported by three groups of workers,[2,4,15] but the amplitude of the oscillations was small, ranging between 10 to 20% of the mean activity.

Five types of changes in enzyme activity in response to D are evident and their relative incidence is summarized in Table 1B. The most frequent response is increase in enzyme synthesis with decreasing D values—either throughout the range of D values employed, or through a substantial part of it, (i.e., when maximum activity is observed at an intermediate D value). This type of response embraces almost all the catabolic enzymes that have been examined (Table 1), except for glycerol dehydrogenase in *Aerobacter aerogenes*,[14] and D-glyceraldehyde- 3-phosphate dehydrogenase in *Erwinia amylovora* 595.[5] Much less frequent is an increase in enzyme activity with increasing D value, and this response appears to be confined primarily to enzymes involved in biosynthetic reactions (NADP-glutamate dehydrogenase, α-acetohydroxy acid synthetase, glutamic-oxalacetic transaminase), and reactions connected with the respiratory chain (NADH oxidase, superoxide dismutase). The other two responses — no change in enzyme activity in response to D, and a minimal activity at an intermediate D value are clearly exceptional.

Figure 1A illustrates the first type of response for three catabolic enzymes of a *Pseu-*

Table 1A
INFLUENCE OF DILUTION RATE ON BACTERIAL ENZYME SYNTHESIS

Enzyme	Organism	Carbon source	Ref.
I. Activity increased with decreasing dilution rate			
Hexokinase	*E. coli*	Glucose	1a
Fructose diphosphate aldolase	*Butyrivibrio fibrisolvens*	Glucose	2
Glucose-6-phosphate dehydrogenase	*Pseudomonas* sp.	L-Lactate or succinate	3
	Spirillum sp.	L-Lactate or succinate	3
Glucose-phosphate isomerase	*Butyrivibrio fibrisolvens*	Glucose	2
Glucose-phosphate isomerase	*Ruminococcus albus*	Glucose	2
Phosphofructokinase	*Butyrivibrio fibrisolvens*		
β-Galactosidase	*E. coli* ML308 (constitutive)	Glucose	4
Phosphoglucomutase	*Butyrivibrio fibrisolvens*	Glucose	2
Pyruvate kinase	*Pseudomonas* sp.	L-Lactate	3
NAD-dependent L-Lactate dehydrogenase	*Pseudomonas* sp.	L-Lactate or succinate	3
	Spirillum sp.	L-Lactate or succinate	3
NAD-independent L-Lactate dehydrogenase	*Pseudomonas* sp.	L-Lactate or succinate	3
	Sprillum sp.	L-Lactate or succinate	3
Aconitase	*Pseudomonas* sp.	L-Lactate	3
Isocitrate dehydrogenase	*Butyrivibrio fibrisolvens*	Glucose	2
	Pseudomonas sp.	L-Lactate or succinate	3
Succinate oxidase	*Erwinia amylovora* 595	Glucose	5
Malate dehydrogenase	*Erwinia amylovora* 595	Glucose	5
Aspartase	*E. coli* B6	Glucose + aspartate	6
Ornithine carbamyl transferase	*Pseudomonas fluorescens*	Citrate	see 7
Amidase	*Pseudomonas aeruginosa* strains L9 & C11 (constitutive)	Succinate + acetamide	8
Peroxidase	*E. coli* K12	Glucose	9
Acid phosphatase	*E. coli* 308	Succinate	see 7
	K. aerogenes	Sucrose	10
II. Activity was maximum at an intermediate dilution rate			
β-Galactosidase	*K. aerogenes*	Lactose	11
	E. coli B6b2	Lactose	11
	E. coli B6	Lactose	11
	E. coli B6	Glucose	11
	E. coli B	Maltose	see 7
	E. coli ML308 (constitutive)	Succinate or glycerol	4
Phosphoglucomutase	*Ruminococcus albus*	Glucose	2
Fructose diphosphate aldolase	*Ruminococcus albus*	Glucose	2
Isocitrate dehydrogenase	*Ruminococcus albus*	Glucose	2
Glutamate dehydrogenase	*Ruminococcus albus*	Glucose	2

Table 1A (continued)
INFLUENCE OF DILUTION RATE ON BACTERIAL ENZYME SYNTHESIS

Enzyme	Organism	Carbon source	Ref.
Aspartate amino-transferase	*Ruminococcus albus*	Glucose	2
Amidase	*Pseudomonas aerugi-nosa* 8602 (wild-type)	Succinate + aceta-mide	8
Acid phosphatase	*K. aerogenes*	Glucose	10
Pullulanase	*K. aerogenes*	Maltose	12
Chondroitinase	*Proteus vulgaris*	Chondroitin sulfate or nicotinic acid	13
Lipase	*Anaerovibrio lypolitica*	Glycerol	see 7

III. Activity increased with increasing dilution rate

D-glyceraldehyde 3-phosphate dehydro-genase	*Erwinia amylovora* 595	Glucose	5
Glycerol dehydrogen-ase	*A. aerogenes*	Glycerol	14
α-Acetohydroxy acid synthetase	*E. coli*	Glucose	1a
Glutamic-oxalacetic transaminase	*E. coli*	Glucose	1a
NADP-linked gluta-mate dehydrogenase	*E. coli*	Glucose	1a
NADH oxidase	*Pseudomonas* sp. *Spirillum* sp.	L-Lactate	3
Superoxide dismutase	*E. coli* K12	Glucose	9
Urease	*Hydrogenomonas* H16	Fructose	see 7

IV. Minimal activity at an intermediate dilution rate

Isocitrate dehydro-genase	*Spirillum* sp.	L-Lactate	3

V. No change in activity in response to dilution rate

2-Oxoglutarate dehy-drogenase	*Pseudomonas* sp.	L-Lactate	3
Succinate dehydro-genase	*Pseudomonas* sp.	L-Lactate	3
Catalase	*E. coli* K12	Glucose	9

domonas sp. grown under L-lactate limitation, i.e., increase in enzyme activity with decreasing D values.[3] Since the K_s for L-lactate of this bacterium is known,[16] it is possible to present enzyme activities as a function of the steady-state L-lactate concentration that would be expected in the culture medium at various D values. Such a plot (Figure 1B) shows that the activity of the three enzymes in the cells decreases by a constant fraction (10 to 60% of the maximal activity) for each doubling of the medium L-lactate concentration; two of the enzymes, isocitrate dehydrogenase and glucose-6-phosphate dehydrogenase, show a slower rate of decrease in activity at higher L-lactate concentrations.

Table 1B
RELATIVE INCIDENCE OF EACH TYPE OF
RESPONSE SUMMARIZED FROM TABLE
1A[a]

Type of response	Number	Percent
Increased activity with decreasing D	24	47
Maximal activity at an intermediate D	15	29
Increased activity with increasing D	8	15
Minimal activity at an intermediate D	1	2
No change	3	5

[a] Total number of responses, 51.

1. Role of Induction and Catabolite Repression

During carbon-limited growth in the chemostat, there is with increasing D a progressive increase in the steady-state concentration of the carbon source at which growth occurs.[17] Thus, for the increase in enzyme activity with increasing D values, the simplest explanation is that the limiting nutrient or its metabolite(s) induces the enzyme synthesis and the induction remains partial at submaximal dilution rates owing to the low concentration of the limiting nutrient in the culture. Similary, the explanation for increased enzyme synthesis with decreasing D values is release of catabolite repression. Such release presumabiy results from decreasing intracellular pools of metabolites as organisms are grown at progressively lower steady-state concentrations of the limiting carbon source. In addition, two lines of evidence suggest that intracellular cyclic AMP (cAMP) levels also play a role in the release of catabolite repression with decreasing D values. First, it appears that the intracellular cAMP concentration of glucose-limited cultures of *Escherichia coli* increases with decreasing D values, [18] which would have the effect of increasing the synthesis of catabolite repressible enzymes, since many of these require cAMP for their synthesis;[19] and second, when cAMP was added to steady state *E. coli* culture, growing at $D = 0.4$ hr^{-1}, there was an increase in β-galactosidase synthesis.[4]

The role of induction and catabolite repression in the increase and decrease of enzyme activity observed under the influence of D has been elegantly demonstrated by Clarke and collaborators, who studied amidase synthesis by *Pseudomonas aeruginosa* and its various regulatory mutants.[8,15] In this organism, amidase is induced by acetamide, but is also subject to repression by products of acetamide metabolism. When *P. aeruginosa* was grown in chemostat under acetamide (or acetamide + succinate) limitation, the organisms exhibited a sharp peak of activity at $D = 0.3$ hr^{-1} (Figure 2). This was interpreted to reflect a balance between induction and catabolite repression : at low D values with low concentration of acetamide in the medium, the degree of induction limited the amount of enzyme synthesized; at D values above 0.3 hr^{-1}, however, increasing concentration of acetamide in the medium led to progressively increased catabolite repression of the synthesis of the enzyme. If this interpretation is correct, mutants constitutive for amidase should exhibit increase in the activity of this enzyme with decreasing D values even below $D = 0.3$ hr^{-1}, since in the absence of a need for induction, catabolite-repression would be the sole mechanism regulating enzyme synthesis. Mutant C11 of this bacterium, which was fully constitutive for amidase, fulfilled this prediction and showed only an increase in enzyme activity as D was decreased to the lowest value examined (Figure 2). Similarly, mutants altered in their sensitivity to catabolite repression would be expected to differ from wild type in the behavior of the enzyme above $D = 0.3$ hr^{-1}, and indeed mutant L11, which was fully

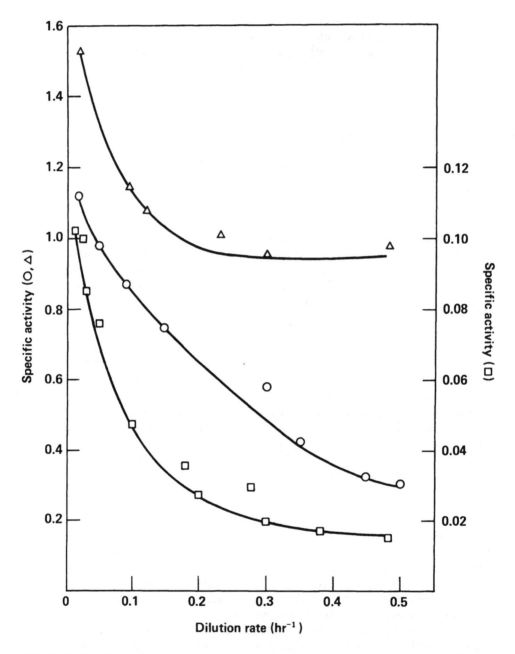

FIGURE 1A Specific activities of various enzymes in a *Pseudomonas* sp. plotted as a function of dilution rate (A), or the steady-state L-lactate concentration in the culture vessel at the various D values (B). The organisms were grown in an L-lactate-limited chemostat; the steady-state concentration of L-lactate in the culture vessel at different D values was calculated from the K, value for L-lactate of this organism reported by Matin and Veldkamp.[16] Symbols: O, ●: NAD-independent L-lactate dehydrogenase; △, ▲: Isocitrate dehydrogenase; □, ■: Glucose-6-phosphate dehydrogenase. Redrawn from the date of Matin et al.[3]

inducible but possessed decreased sensitivity to catabolite repression, showed an onset of induction with D similar to the wild type but a much more gradual decline in activity with increasing D value (Figure 2).

Further evidence that decrease in the activity of an inducible enzyme at low D values is the result of decreasing inducer concentration was provided by Smith.[20] β-galactosid-

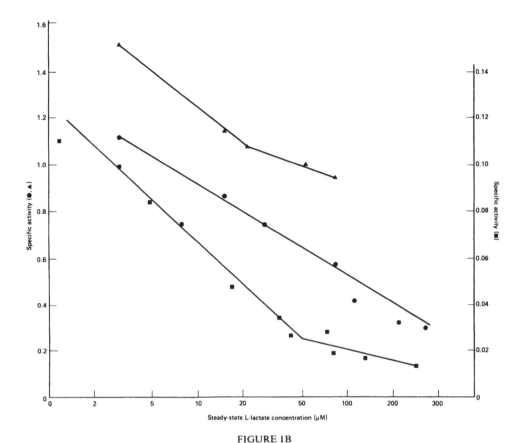

FIGURE 1B

ase activity in *Aerobacter (Klebsiella) aerogenes* shows maximum activity at intermediate D values,[21] and Smith reasoned that if the decrease in the enzyme activity below the inflection point is due to decreasing induction, then it should not occur if the inducer concentration were maintained constant, irrespective of D. He realized this condition by including in the medium a nonmetabolizable inducer for β-galactosidase (TMG); under these conditions maximum enzyme activity was observed at the lowest D value examined (Figure 3).

These studies strongly support the view that induction and catabolite repression are the mechanisms for many of the changes in enzyme levels that occur in response to culture D value. In at least one instance, however, lack of sufficient inducer concentration was clearly not the mechanism for decreased enzyme activity at low D values. McLeod et al.[4] found that *E. coli* ML 308, which is constitutive for β-galactosidase, showed maximum activity for this enzyme at D = 0.20 to 0.24 hr^{-1} when grown under succinate or glycerol limtation. They ascribed the decrease in enzyme activity at D values below the inflection point to scarcity in the cells of precursors required for the synthesis of the enzyme.

2. Identity of Metabolites Regulating Enzyme Synthesis

Despite a great deal of effort,[22] the identity of metabolites controlling synthesis of individual enzymes remains uncertain. There is good evidence that intracellular cAMP levels play an important role in mediating catabolite repression of β-galactosidase,[23]

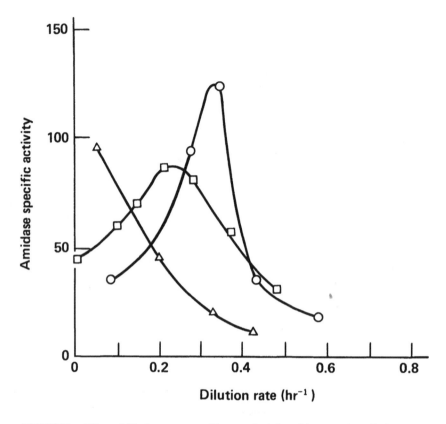

FIGURE 2.   Effect of dilution rate on amidase synthesis by wild type and regulatory mutants of *Pseudomonas aeruginosa* grown in continuous culture under steady-state conditions in minimal medium containing 10 mM succinate + 20 mM acetamide. (Succinate was included to amplify catabolite repression.) ○, wild type; △, mutant C11 (fully constitutive for amidase); □, mutant L11 (fully inducible but less sensitive to catabolite repression). Rearranged from Clarke et al.[8]

but even for this intensively studied enzyme it is not known what other metabolites may be involved in this regulation. A good deal of confusion in the literature dealing with enzyme regulation has resulted from the use of batch cultures in these studies: the continually changing environment of such cultures and the fact that enzyme synthesis rates can be altered only by qualitative changes in the growth environment make such studies irreproducible and difficult to interpret. The use of continuous culture technique in these studies holds a special promise. As we have seen (Table 1), large changes in the rate of enzyme synthesis of the cells can be achieved without any qualitative change in the environment and under steady-state conditions. Morever, by employing different nutrient limitations, one can manipulate the concentration of different types of intracellular metabolites, thus affording a facile and valuable means for obtaining information on the nature of metabolites involved in the regulation of the synthesis of individual enzymes. With changing D values under the limitation of a given nutrient, the concentration of metabolites derived from that nutrient will tend to change selectivity, so that the behavior of an enzyme to changing D values under several individual limitations can provide information about the nature of the molecules regulating its synthesis. For instance, Matin et al.[3] found that in a *Pseudomonas* sp., the activity of five catbolic enzymes (NAD-independent and NAD-dependent L-lactate dehydrogenases, aconitase, isocitrate dehydrogenase, and glucose-6-phosphate dehydrogenase) increased with decreasing D—not only under carbon limitation, but

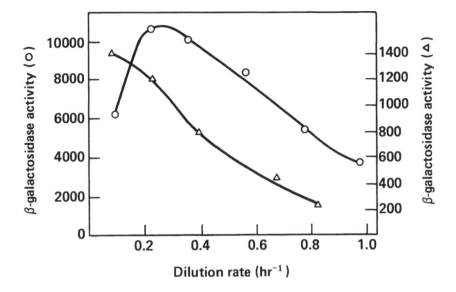

FIGURE 3. The effect of dilution rate on the stead-state levels of β-galactosidase specific activity of *Klebsiella aerogenes*, grown in lactose-limited chemostate (O), or maltose-limited chemostat induced with TMG ($5 \times 10^{-4} M$) (Δ). Redrawn from Dean,[7] and Smith and Dean.[21]

also under that of NH_4^+, or phosphorous, and concluded that the catabolite repressor molecules for these enzymes were probably compounds of carbon, nitrogen, and phosphorous; and results of a similar approach were consistent with the conclusion that N-acetyl-D-glucosamine-6-phosphate is involved in the repression of alkaline and neutral proteases in *Bacillus subtilis*.[24] Such studies can conceivably provide reliable information on the identity of metabolites involved in the regulation of enzyme synthesis. One can grow a bacterium to a steady state in a chemostat at various D values under different individual limitations and correlate the levels of a given enzyme with those of certain selected intracellular metabolites to determine which ones exhibit parallel changes. The findings of such experiments can be extended by repeating the measurements with mutants of the bacterium altered in their capacity to synthesize the relevant metabolite(s) and/or the enzyme.

3. Adaptive Significance of Enzyme Responses

Since the most frequent response to decreasing concentration of carbon compounds in the growth environment is increased enzyme synthesis, the potential benefits of this response will first be discussed. It seems *a priori* paradoxical that slower growing cells should possess higher levels of catabolic enzymes and in fact runs counter to the conclusions based on earlier batch culture work. However, its potential beneficial effect is easy to infer. Since the K_m of many catabolic enzymes lies in mM range, it follows that in the presence of μM concentrations of carbon compounds (as happens in carbon-limited chemostat cultures of bacteria, especially at low D vaues), increased enzyme levels would enhance the effectiveness of the bacteria in metabolizing these substrates. That higher levels of a catabolic enzyme are indeed of selective advantage at low concentration of carbon compounds is suggested by the work of Horiuchi et al.,[26] who showed that at low D values under lactose limitation, the selection of mutants of *E. coli*, which are constitutive for β-galactosidase, is favored. As opposed to the inducible wild type, in which β-galactosidase levels would depend on the concentration of lactose outside the cell and would therefore be quite low at low D values, in the constitutive strain they would be maximal irrespective of the D value, and this difference probably accounted for the selective advantage of the constitutive strain at low D values.[26] This conclusion is strengthened by the finding that prolonged cultivation at low contration of carbon compounds, (i.e., at low D values in carbon-limited chemostat) leads to the selection of mutants containing markedly elevated levels of the "substrate-capturing" enzyme, i.e., the first enzyme involved in the metabolism of the carbon substrate used. Thus, "hyper" mutants of *E. coli,* containing several fold higher levels of β-galactosidase than the normal constitutive strain, are selected during prolonged cultivation at low D values under lactose limitation;[26] and during xylulose-limited growth, *K. aerogenes* strains are selected that contain elevated levels of ribitol dehydrogenase, the first enzyme of xylulose catabolism in this organisms.[27, 27a] The work of Horiuchi et al.[26] also suggests that environments of low nutritional status would select for constitutive over inducible strains. It may be pertinent in this regard that the two fresh water bacteria isolated by Matin and Veldkamp by continuous culture enrichment, and maintained in culture without ever being exposed to high concentration of any carbon compound, showed a pattern of enzyme synthesis in response to D, which was consistent with the idea that most of their catabolic enzymes examined were constitutive, i.e., they required no induction and showed only increased catabolite repression in response to increasing D.[3,16]

Another potential benefit of a general release of catabolite repression in environments containing low concentration of carbon compounds is that it would prime the organisms to utilize new substrates without a lag, should they bcome available, and would enable them to utilize more than one substrate concurrently rather than exhibiting the diauxic or biphasic pattern as is commonly found at high nutrient concentrations. Both these effects have obvious survival value in nutritionally poor environments, and the latter is discussed in greater detail below.

If increased enzyme levels are beneficial at low D values, as has been suggested above, what is the rationale of decreased levels of biosynthetic enzymes at lower D values (Table 1)? The answer probably relates to the lack of necessity of maintaining a rapid flux of metabolites through anabolic sequences, as opposed to the catabolic sequences, and it must be assumed that the lowered enzyme levels can adequately handle the reduced biosynthetic needs at lower growth rates. The reduced activity of biosynthetic enzymes at low D values probably contributes to the flux of metabolites through the catabolic sequences by lowering the drainage of metabolites from these sequences.

4. Cofactor Concentration and Ratios

With decreasing D values under carbon limitation, there was a logarithmic increase in the NAD(H), (i.e., NAD + NADH) content of a *Pseudomonas* sp. (Figure 4),[28] although the NADH content remained invariant. The NAD(H) content was strictly a function of culture D and was independent of the nature of the carbon source employed (Figure 4). The increased NAD content probably facilitated the metabolism of the limiting carbon substrate supplied at progressively lower concentration with decreasing D, as do, evidently, the increased levels of catabolic enzymes (see above). Another probable function of this increase concerns the cellular NADH concentration and the apparent need of the bacterial cell to maintain it within a narrow concentration range under various growth conditions.[28,29] As pointed out above, with decreasing D under carbon limitation, bacteria can be expected to have progressively lower concentration of intracellular metabolites including those whose oxidation by NAD-linked dehydrogenases serves as the source of NADH. A concomitant increase in oxidized NAD cofactor would tend to compensate for this lowered concentration and thereby

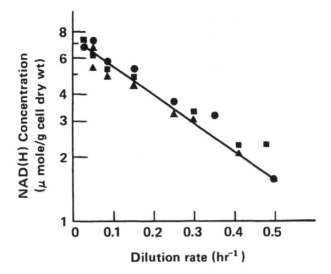

FIGURE 4. The effect of dilution rate on NAD(H) (i.e., NAD + NADH) concentration of cells of *Pseudomonas* sp. in carbon-limited media. ●, L-lactate-limited medium; ▲, succinate-limited medium; ■, glucose-limited medium. (From Matin, A. and Gottschall, J. C., *J. Gen. Microbiol.*, 94, 333, 1976. With permission.)

help maintain a constant NADH pool size. This interpretation was supported by the finding that bacteria grown under NH_4^+ limitation, which would be expected to posses an excess of carbon metabolites at all D values, did not exhibit an increase in NAD concentration with decreasing D values.

Dolezal and Kapralek have examined the effect of D on the content of intracellular adenylate nucleotides and energy charge during carbon- and ammonium-limited growth.[30] Up to D values equaling ∿60% of μ_{max}, the concentration of ATP, ADP, and AMP was independent of the culture D value, and the energy charge remained poised at ∿0.92. Above this D value, the nucleotide concentration increased, but the energy charge remained unaffected. These results are consistent with the idea that the relative adenylate concentration, (i.e., the energy charge) has an important regulatory function and must be rigidly maintained with a narrow concentration range during logarithmic growth.[31]

B. Eukaryotes

The influence of D on enzyme activity of eukaryotes, such as yeast, fungal, and plant cells, grown in a chemostat under carbon limitation, is summarized in Tables 2A and 2B. As in the case of bacteria (Tables 1A and 1B), five types of response are evident and it is a safe assumption that induction and catabolite repression play a role also in eukaryotes in determining these responses. However, the relative incidence of individual types of responses is markedly different in eukaryotes from that observed in bacteria. In contrast to the latter, increased enzyme synthesis in eukaryotic cells is much more frequently observed with increasing rather than decreasing D values: among the examples listed, this was the situation in 49% of cases, whereas in only 27% of instances increased enzyme activity accompanied a decrease in D value. Similarly, the absence of any effect of culture D value on the levels of enzymes was much more frequently observed in eukaryotes than in bacteria. It must be assumed that the differences in the cellular organization of these two groups of organisms necessitates different strategies to meet the demands of environments of different nutritional status.

Table 2A
INFLUENCE OF DILUTION RATE ON EUKARYOTIC
ENZYME SYNTHESIS

Enzyme	Organism	Carbon source	Ref.
I. Activity increased with decreasing dilution rate			
Glucose-6-phosphate dehydrogenase	*Candida utilis*	Sucrose	32
Fructose diphosphatase	*Aspergillus niger*	Glucose	33
Isocitrate lyase	*Aspergillus niger*	Citrate	33
Cellulase	*Trichoderma viride*	Glucose	34
Esterase	*Aspergillus niger*	Citrate	33
II. Activity was maximum at an intermediate dilution rate			
Phosphofructokinase	*Acar pseudoplatanus L*	Sucrose	35
Malate dehydrogenase	*Saccharomyces cerevisiae*	Glucose	36
Isocitrate lyase	*Saccharomyces cerevisiae*	Glucose	36
Malate synthase	*Saccharomyces cerevisiae*	Glucose	36
	Aspergillus niger	Glucose or citrate	33
Succinate cytochrome c reductase	*Saccharomyces cerevisiae*	Glucose	36
NAD-linked Glutamate dehydrogenase	*Saccharomyces cerevisiae*	Glucose	36
Invertase	*Saccharomyces carlsbergensis*	Glucose	37
III. Activity increased with increasing dilution rate			
Hexokinase	*Aspergillus niger*	Glucose	33
	Aspergillus nidulans	Glucose	38
Glucose-6-phosphate dehydrogenase	*Aspergillus niger*	Glucose or citrate	33
	Aspergillus nidulans	Glucose	38
	Acar pseudoplatanus L	Sucrose	35
Phosphogluconate dehydrogenase	*Aspergillus niger*	Glucose	33
	Acar pseudoplatanus L	Sucrose	35
Phosphofructokinase	*Aspergillus niger*	Glucose	33
Fructose diphosphate aldolase	*Aspergillus nidulans*	Glucose	38
	Saccharomyces cerivisiae	Glucose	36
Pyruvate kinase	*Acar pseudoplatanus L*	Sucrose	35
Citrate synthase	*Aspergillus niger*	Glucose	33
Aconitase	*Aspergillus niger*	Glucose or citrate	33

Table 2A (continued)
INFLUENCE OF DILUTION RATE ON EUKARYOTIC
ENZYME SYNTHESIS

Enzyme	Organism	Carbon source	Ref.
Isocitrate dehydro-genase	*Aspergillus niger*	Glucose or citrate	33
	Candida utilis	Sucrose	32
Isocitrate dehydro-genase-NAD-linked	*Aspergillus niger*	Citrate	32
Malate dehydrogen-ase	*Aspergillus niger*	Glucose or citrate	33
NADP-Glutamate dehydrogenase	*Saccharomyces cerevisiae*	Glucose	36
	Aspergillus nidulans	Glucose	38
Tyrosinase	*Aspergillus nidulans*	Glucose	38
3-Carboxymuconate cyclase	*Aspergillus nidulans*	β-Hydroxyben-zoate	39
3-Carboxymucono-lactone hydrolase	*Aspergillus nidulans*	β-Hydroxyben-zoate	39

IV. Minimal activity at an intermediate dilution rate

Hexokinase	*Candida utilis*	Sucrose	32

V. No change in activity in response to dilution rate

Hexokinase	*Aspergillus niger*	Citrate	33
Glucose-6-phophate dehydrogeanse	*Rhodotorula mucilaginosa*	β-Hydroxyben-zoate	39
Phosphogluconate dehydrogenase	*Aspergillus niger*	Citrate	33
Phosphofructokinase	*Aspergillus niger*	Citrate	33
Fructose diphospha-tase	*Aspergillus niger*	Citrate	33
Pyruvate kinase	*Aspergillus niger*	Glucose or citrate	33
Citrate synthase	*Aspergillus niger*	Citrate	33
Isocitrate dehydro-genase-NAD-linked	*Aspergillus niger*	Glucose	32
Esterase	*Aspergillus niger*	Glucose	33
4-Hydroxy-benzoate 3-monoxygenase	*Rhodotorula mucilaginosa*	β-Hydroxyben-zoate	39
Protocatechuate 3,4-dioxygenase	*Rhodotorula mucilaginosa*	β-Hydroxyben-zoate	39

III. QUANTITATIVE RELATIONSHIP BETWEEN SUBSTRATE CONCENTRATION AND ENZYME SYNTHESIS

Satisfactory exploration of the quantitative relationship between substrate concentration and enzyme synthesis requires that the concentration of the substrate be held constant at fixed values during the growth of the organism. This condition is very adequately met by employing the continuous culture technique and many studies have taken advantage of this fact.

Table 2B
RELATIVE INCIDENCE OF EACH TYPE OF
RESPONSE SUMMARIZED FROM TABLE 2A*

Type of Response	Number	Percent
Increased activity with decreasing D	6	12
Maximum activity at an intermediate D	7	15
Increased activity with increasing D	23	49
Minimal activity at an intermediate D	1	2
No change in activity in response to D	11	23

* Total number of responses, 47

A. Glucose Metabolism in *Pseudomonas aeruginosa*

Glucose metabolism in *P. aeruginosa* is complex and can occur by the "intracelluar phosphorylation pathway", involving the direct uptake of glucose by a glucose transport system followed by intracellular phosphorylation, and /or the "direct oxidative pathway", which involves extracellular oxidation of glucose to gluconate and 2-oxogluconate, the uptake of these compounds by specific transport systems, and their intracellular phosphorylation (Figure 5). Many of the enzymes involved in glucose metabolism are induced by glucose, but are subject to repression by citrate, and Dawes and collaborators have made a throrough study of the concentration effect of these substrates on glucose metabolism of this bacterium[40,41]

P. aeroginosa strain PAO1 was grown with increasing concentration of glucose in the inflow medium, which contained 75 mM citrate and NH_4^+ as the growth-limiting nutrient. Organisms grown on citrate alone contained low specific activities of the transport system for glucose, gluconate, and 2-oxogluconate (Figure 6). Inclusion of glucose in the medium made little difference until the glucose concentration reached 6 to 8 mM whereupon a doubling in the glucose transport activity was observed; and the activity continued to increase up to a glucose concentration of 15 mM, but further increase in this concentration decreased the activity (Figure 6). In contrast, the transport systems for gluconate and 2-oxogluconate showed significant increases only at glucose concentrations above 15 mM. At a glucose concentration of 45 mM, a decrease in citrate concentration from 75 to 45 mM increased the activity of the transport systems for gluconate and 2-oxogluconate (Figure 6).

Significant amounts of gluconate, generated by the extracellular oxidative pathway, were found in the culture fluid at higher glucose concentrations in the inflow medium. That this gluconate led to the induction of the transport system for gluconate, and repression of that for glucose at 15 mM and higher glucose concentrations was shown using a mutant that lacked glucose dehydrogenase: such a mutant inhibited neither the induction of the gluconate transport system nor the repression of that for glucose up to a glucose concentration of 45 mM in the inflow medium.

The effect of changes of glucose and citrate concentrations during growth in an NH_4^+-limited medium on glucose catabolic enzymes was similar to that discussed above for the glucose transport system (Figures 7 and 8). Citrate-grown cells possessed little activity, but once a glucose concentration of 6 to 8 mM was attained, there was significant induction; further, at a fixed glucose concentration, the activity of hexokinase and glucose-6-phosphate and gluconate dehydrogenases could be increased by omitting citrate from the medium (Figures 7 and 8). This apparent repression by citrate of glucose catabolic enzymes was investigated in greater detail using another strain of *P. aeruginosa* (2F32). Again an NH_4^+-limited inflow medium was employed, but the glucose concentration was kept constant and that of citrate varied. There was a progres-

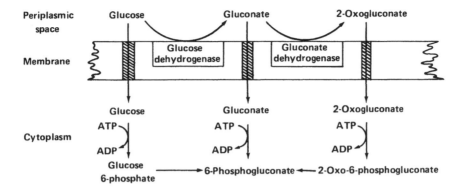

FIGURE 5. Extracellular and intracellular pathways of glucose catabolism and related transport systems in *Pseudomonas aeruginosa*. (From Whiting, W. H., Midgley, M., and Dawes, E., *Biochem. J.*, 154, 659, 1976. With permission.)

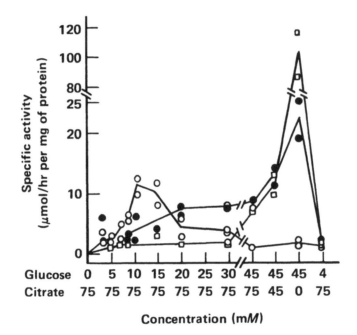

FIGURE 6. Effect of various concentrations of citrate and glucose in the inflowing medium on the transport systems for glucose, gluconate, and 2-oxogluconate in NH₄⁺-limited *Pseudomonas aeruginosa* PA01. Transport systems for glucose, O; gluconate, ●; 2-oxogluconate, □. (From Whiting, W. H., Midgley, M., and Dawes, E., *Biochem. J.*, 154, 659, 1976. With permission.)

sively more pronounced repression of the glucose enzymes as the citrate concentration of the inflow medium was increased and, as was the case with the inductive effect of glucose, a threshold citrate concentration of 8 to 16 m*M* had to be reached before the observed repression became significant (Table 3).

These results are consistent with the ideas that glucose enzymes in *P. aeruginosa* are induced by glucose or its metabolites and repressed by citrate or the products of is degradation, and indicate that the effectiveness with which the inducer and the repressor compete with each other for exerting their respective influences is highly concentration dependent. The molecular basis of this competition remains unknown.

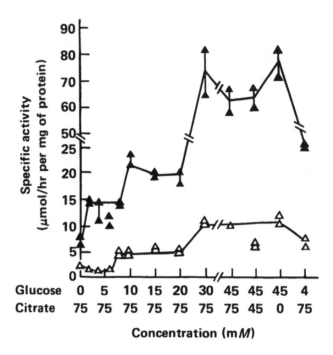

FIGURE 7.   Effect of various concentrations of citrate and glucose in the inflowing medium on the enzymes of the extracellular pathway of glucose catabolism of NH₄⁺-limited *Pseudomonas aeruginosa* PA01. Glucose dehydrogenase, △; gluconate dehydrogenase, ▲. (From Whiting, W. H., Midgley, M., and Dawes, E., *Biochem. J.*, 154, 659, 1976. With permission.)

B. Operation of Fermentative and Oxidative Routes in Facultative Anaerobes

Oxygen tension plays a critical role in the expression of fermentative and oxidative metabolism of facultative anaerobes, and this role has been studied with cells grown in chemostats under steady-state conditions of defined partial pressures of oxygen. For an *E. coli* culture growing in glucose-mineral salts medium at D = 0.2 hr⁻¹, the dissolved oxygen tension became zero at an input p0₂ of 30 mm Hg, and changes in input p0₂ around this value led to pronounced effects on the enzymic composition of the organisms.[42] Reduction of the input p02 below 50 to 30 mm Hg resulted in a progressive increase in glucose-6-phosphate and 6-phosphogluconate dehydrogenases, aldolase, and phosphofructokinase activities (Figure 9). Similar increases in the activities of fermentative enzymes or total fermentative potential under these conditions have been demonstrated in *A. nidulans*, *Erwinina amylovora*, and *S. cerevisiae*.[5,38,43] In the *E. coli* culture, the activity of the components of the respiratory chain showed a complex pattern of change in response to oxygen tension (Figure 10).[42] Between p0₂ values of about 30 and 15 mm Hg, NADH oxidase, succinate dehydrogenase, and cytochromes a₂ and b showed increasing activity with decreasing p0₂, but further decrease in p0₂ led to a rapid decrease in these activities. Isocitrate and α-ketoglutarate dehydrogenases, in contrast, showed a linear decrease in activity with decreasing input p0₂, the latter enzyme becoming undetectable below an input pO₂ of 30 mm Hg. A similar pattern of change in the activity of respiratory enzyme at different oxygen tensions has been reported in *S. cerevisiae*, *A. aerogenes*, *E. amylovora,* and *Mucor gerevensis*, except that in most of these instances the Kreb cycle enzyme activities also showed maxima at intermediate p0₂ values.[5,44-46]

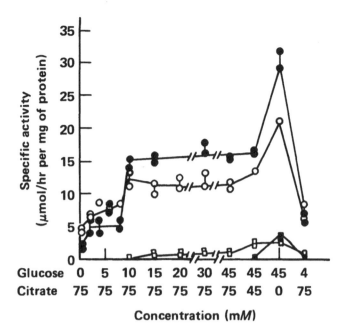

FIGURE 8. Effect of various concentrations of citrate and glucose in the inflowing medium on the enzymes of the intracellular pathway of glucose catabolism of NH_4^+-limited *Pseudomonas aeruginosa* PA01. Hexokinase, O; glucose 6-phosphate dehydrogenase, ●; gluconate kinase, □; 2-oxogluconate enzymes, ■. (From Whiting, W. H., Midgley, M., and Dawes, E., *Biochem. J.*, 154, 659, 1976. With permission.)

These results indicate that under fully aerobic conditions the Krebs cycle is maximally induced and the metabolism is entirely oxidative. Since the activities of glucose-6-phosphate and 6-phosphogluconate dehyrogenases are higher under these conditions than those of aldolase and phosphofructokinase, it appears that the hexose monophosphate (rather than the Embeden- Meyerhof pathway) is the major primary route of glucose catabolism under these conditions. As the decreasing oxygen tensions force the metabolism to become more and more fermentative, there is cessation of the synthesis of superfluous respiratory enzymes; moreover, the greater increases in phosphofructokinase and aldolase activities as compared to the two ezymes of the hexose monophosphate pathway under these conditions suggest that with the increase in anaerbiosis, Embdin-Meyerof pathway assumes a more important role in glucose catabolism[42] The initial increase in the activity of respiratory enzymes with decreasing pO_2 probably represents an attempt on the part of the organisms to make the most of the dwindling oxygen supply. The molecular mechanism of the regulation of enzyme synthesis by molecular oxygen remains unknown, and as with the modulation of enzyme synthesis by D(see above), continuous culture offers a convenient system for investigating it.

C. Mechanism of Pasteur Effect

Passonneau and Lowry postuated that Pasteur effect, i.e., inhibition of the Embden-Meyerhof pathway by oxygen resulted from allosteric inhibition of phosphofructokinase activity by ATP: ATP content of the cells increased under aerobic conditions, which inhibited phosphofructokinase activity and hence glycolysis.[47] This postulate was not supported by later studies, which showed that the steady-state ATP level in

Table 3

EFFECT OF RELATIVE GLUCOSE: CITRATE CONCENTRATIONS IN INFLOWING MEDIUM ON THE STEADY-STATE ENZYMIC SPECIFIC ACTIVITIES OF GLUCOSE-GROWN CELLS

Glucose: citrate contentration in medium (mM)	Specific activity (μmol/hr/mg of N)						
	Glucose 6-phosphate dehydrogenase	Hexokinase	6-Phosphogluconte dehydrogenase	Entner-Doudoroff enzymes	Glucose dehydrogenase	Gluconate dehydrogenase	Isocitrate dehydrogenase
45:0	110.1	22.4	19.8	27.9	17.7	292.6	143.7
45:4	107.0	21.8	19.5	22.9	11.2	312.3	147.0
45:8	119.6	21.7	21.9	28.1	14.0	301.9	176.7
45:16	72.b	19.9	13.0	13.9	14.0	211.2	181.0
45:32	76.9	16.6	11.2	8.7	11.2	198.7	176.6
45:64	40.1	12.5	2.2	3.0	4.4	46.3	180.7
0:64	5.8	3.2	1.0	0.6	4.5	13.9	174.0

Note: The dilution rate was 0.25 hr^{-1}. The values recorded are for the steady states established at each concentration ratio. Reproduced from Ng, F.M.W. and Dawes, E. A., *Biochem. J.*, 132, 129, 1973. With permission.

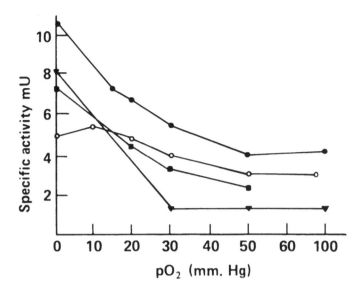

FIGURE 9. Effect of input pO₂ on the activities of the glucose metabolizing enzymes of *E. coli* K-12. Symbols: ●, glucose-6-phosphate dehydrogenase; ○, 6-phosphogluconate dehydrogenase; ■, Fructose diphosphate aldolase; ▼, Phosphofructokinase. (From Thomas, A. D., Doelle, H. W., Westwood, A. W., and Gordon, G. L., *J. Bacteriol.*, 112, 1099, 1972. With permission.)

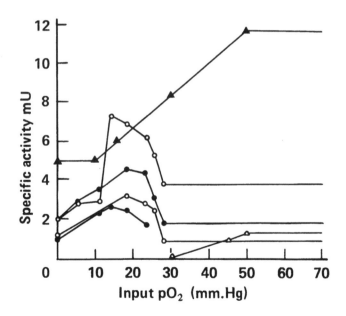

FIGURE 10. Effect of input pO₂ on the respiratory activity of *E. coli* K-12. Symbols: ○, NADH oxidase; ●, cytochrome a₂; □, cytochrome b₁; ■, succinate dehydrogenase; ▲, isocitrate dehydrogenase; △, 2-oxoglutarate dehydrogenase. (From Thomas, A. D., Doelle, H. W., Westwood, A. W., and Gordon, G. L., *J. Bacteriol.*, 112, 1099, 1972. With permission.)

aerobic and anaerobic cells was quite similar.[48] Further, Thomas et al. demonstrated that unlike the pophofructokinase synthesized under anaerobic conditions, the aerobic phosphofructokinase was not allosterically inhibited by ATP. [42] These observations, as well as changes in the activity of fermentative enzymes in response to oxygen tension (see above), or glucose concentration, led to the following proposal for the mechanism of Pasteur effect.[42] Although only low levels of phosphofructokinase are present in the cells under aerobic conditions, the enzyme can still function significantly, since it is not inhibited by ATP. With aldolase activity being low, fructose diphosphate would tend to accumulate, which would inhibit 6-phosphogluconate dehydrogenase activity. Thus phosphofructokinase activity is not a limiting factor in the control of aerobic glucose metabolism and it is conceivable that aldolase and glucose-6-phosphate dehydrogenase activities are involved.

IV. MIXED SUBSTRATE UTILIZATION

It is well established that when presented with more than one carbon and energy substrate under the nutrient-sufficient conditions of batch culture, bacteria usually metabolize only the more readily utilizable substrate, while enzymes for the dissimilation of the other remain repressed. Such a preference for one compound to the exclusion of the rest would be of little value in a nutritionally poor environment, and the selective pressure would favor those organisms capable of making use of different substrates simultaneously. There is increasing evidence that bacteria are, in fact, capable of utilizing multiple substrates concurrently during growth at low nutrient concentrations. The first clear evidence of this was provided by Mateles et al.,[49] who studied the growth of *E. coli* and *P. fluorescens* in carbon-limited media fed both glucose and fructose, substrates which were not simultaneously metabolized by this organism in batch culture. Both the bacteria exhibited concurrent utilization of the two sugars at low D values, as evidenced by the fact that the residual amount of the two sugars in steady-state cultures at these D values was very low (Figure 11); at higher D values, i.e., at higher concentration of the sugars in the culture, only glucose was utilized by both the bacteria.[49] The influence of D on this phenomenon was more pronounced with the pseudomonad, as in this case the D had to be lowered to very low values before the concurrent utilization of the two sugars could occur. Similary, an inducible strain of *E. coli* utilized glucose and lactose simultaneously in a carbon-limited continuous culture at low D values, but not at high D values; evidently, the utilization of lactose along with glucose at low values was made possible not only by the release of catabolite repression of β-galactosidase synthesis, but also by the lack of interference by low glucose concentrations with the uptake of the inducer for this enzyme (i.e., lactose). [6,11] A mixture as aspartate and glucose was also simultaneously utilized by *E. coli* at low D values, but not high D values.[6]

Recent studies in my laboratory have extended these findings to the utilization of organic and inorganic energy substrates by the mixotrophic chemolithotroph, *Thiobacillus novellus*. This organism can grow heterotrophically in a glucose medium (8 hr generation time) and autotrophically in a thiosulfate medium (20 hr generation time) (Figure 12 A and B). In the presence of a mixture of glucose and thiosulfate, the growth rate and the rate of glucose utilization are reduced compared to the heterotrophic medium, and although both thiosulfate and glucose are concomitantly utilized and eventually disappear from the medium completely, there is no increase in growth yield in this medium over that observed at the corresponding concentration of glucose alone (Figure 12). Thus, although under conditions of nutrient excess this bacterium can utilize the two substrates simultaneously, it derives no benefit from this capacity and is, in fact, hampered by it. [53] Repression in the mixotrophic medium of enzymes of

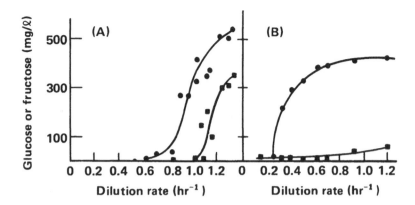

FIGURE 11. Concentration of residual sugars in steady-state carbon-limited cultures of *E. coli* (A), and *P. fluorescens* (B) supplied with a mixed feed of glucose and fructose. ●, residual fructose; ■, residual glucose. (From Mateles, R. I., Chian, S. K., and Silver, R., *Microbial Physiology and Continuous Culture,* Powell, E. O., Evans, C. G. T., Strange, R. E., and Tempest, D. W., Eds., Her Majesty's Stationary Office, London, 1967, 232. With permission.)

glucose metabolism and the glucose transport system, as well as inhibition of these enzyme systems by thiosulfate and sulfite apparently account for the reduced rate of glucose metabolism in such media.[53,54] During nutrient-limited growth, in contrast, concurrent utilization of thiosulfate and glucose is evidently beneficial. This was indicated by chemostat experiments in which this organism was grown under the dual limitation of glucose and thiosulfate. Under these conditions the steady-state biomass was twofold higher than when the inflow medium contained glucose alone at the corresponding concentration (Table 4).[55] Similarly, in *Pseudomonas oxalaticus,* which can obtain cell carbon heterotrophically from organic compounds, such as oxalate, or autotriphically from formate, (i.e., CO_2), either one or the other mode of carbon assimilation prevails in batch cultures. However, when this bacterium was grown in a chemostat under carbon limitation with a combined feed of oxalate and formate, both the heterotrophic and autotrophic pathways of carbon assimilation operated simultaneously at appropriate oxalatermate ratios. This was evidenced by the complete utilization of the two substrates under these conditions and the pattern of synthesis of the enzymes involved in their assimilation (Figure 13).[49a] As noted above, such concurrent utilization of a mixture of substrates during nutrient-limited growth would appear to be an additional advantage of the tendency toward diminished catabolite repression under these conditions.

V. SPECIAL PATHWAYS AND "ISOENZYMES" SYNTHESIS

Several instances are known of the use of special pathways by bacteria to meet the demands of nutrient-limited environments. In *A. aerogenes* grown with excess NH_4^+, the NH_4^+-assimilating reaction is catalyzed by glutamate dehydrogenase (GDH) in which 2-oxoglutarate acts as the NH_4^+ acceptor molecule. However, organisms grown under NH_4^+-limitation in a chemostat possess very low levels of this enzyme, and assimilate NH_4^+ by a different mechanism. This involves the combined functioning of of glutamine synthetase (GS) and glutamine-oxoglutarate amidotransferase, which catalyzes reductive transfer of amide nitrogen to 2-oxoglutarate. This change appears to be an adaptation to low concentration of NH_4^+ in the environment under these conditions, since the K_m of GDH for N_4^+ is $\simeq 10$ mM, and that of GS is $\simeq 0.5$ mM, and the latter enzyme is therefore more suitable for scavenging NH_4^+.[50]

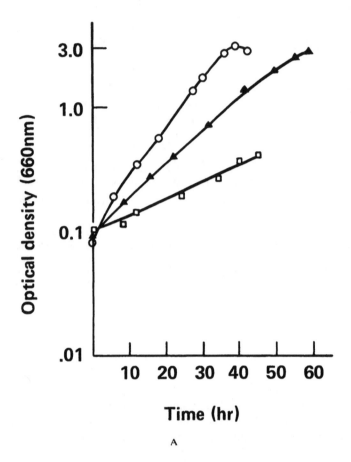

A

FIGURE 12.   Growth parameters of *T. novellus* in various media.
The initial substrate concentrations were: glucose, 0.4%; thiosulfate,
1%. The mineral salts base used was supplemented with 0.03% yeast
extract to satisfy the requirement for organic sulfur compounds.[25] The
experiments were carried out in a New Brunswick Bioflo fermentor
and the pH was automatically maintained at 6.9 ± 0.1 by a pH stat,
employing 5% Na_2CO_3 solution. Growth was determined by turbidity
measurements; one O.D. unit equals approximately 50 mg (dry
weight) of cells per 100 mℓ. A. Growth in autotrophic (thiosulfate,
□), heterotrophic (glucose, ●), and mixotrophic (thiosulfate-glucose,
▲) media. B. Substrate utilization. O, ▲; glucose utilization in heterotrophic and mixotrophic media, respectively; ▼, thiosulfate utilization in mixotrophic medium. (Redrawn from the data of Perez, R. C.
and Matin, A., *J. Bacteriol.*, 142, 633, 1980.)

A. aerogenes possess dual pathways for the assimilation of glycerol as well, one
involving a glycerol kinase, and the other a glycerol dehydrogenase.[51] The K_m for glycerol of the former enzyme is 1-2 × 10⁻⁶ M, whereas that of the latter enzyme is 2-4 ×
10⁻² M, and these pathways therefore represent high and low affinity routes for glycerol
metabolism. Neijssel et al. showed that during glycerol-limited growth under aerobic
conditions, the organisms utilized the glycerol kinase pathway of glycerol metabolism
, but under conditions of glycerol excess, such as prevail in NH_4^+- or phosphate-limited
cultures, they utilized the glycerol dehydrogenase pathway. The advantage derived
from a shift to the high affinity route under glycerol limitation is obvious; the shift to
the low affinity route under conditions of glycerol excess probably serves to protect

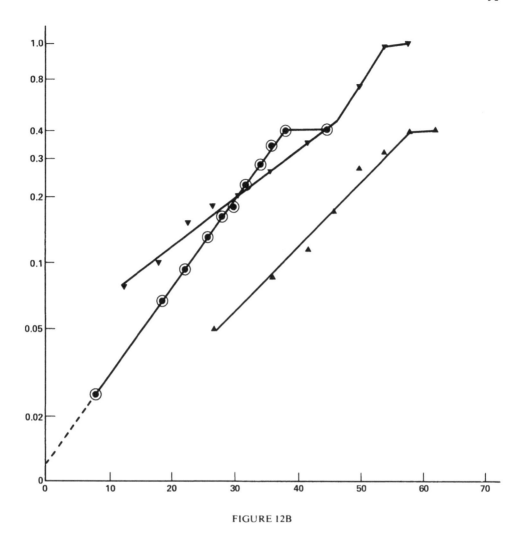

FIGURE 12B

the cell from the buildup of excessive amounts of potentially toxic metabolites from glycerol.[51]

A precisely analogous situation obtains in *P. aeruginosa* with respect to glucose metabolism. As has been pointed out already (Figure 5), this organism possesses two routes for glucose metabolism, the intracellular phosphorylative pathway and the extracellular direct oxidative pathway.[41] When grown in the presence of an excess of glucose, as for instance under NH_4^+-limitation, the organism contained a very low glucose transport activity, but high levels of gluconate and 2-oxogluconate transport systems, which suggests that under these conditions the extracellular direct oxidation pathway operated. Upon transfer to glucose-limited conditions, the activity of the glucose transport system increased fivefold, while that of the transport systems for gluconate and 2-oxogluconate decline five to sevenfold.[52] The Km of the glucose transport system for glucose is 8×10^{-6} *M*, whereas the K_m values for gluconate and 2-oxogluconate uptake are nearly 10^{-3} *M* and it may therefore be surmised that the changeover to the intracellular phosphorylation pathway under glucose limitation ensures scavenging of the glucose present at low concentrations.[52]

These examples suggest that the existence of multiple pathways for performing specific metabolic tasks may be a widespread phenomenon. Such duplicity of functions

Table 4

GROWTH PARAMETERS OF *THIOBACILLUS NOVELLUS* IN NUTRIENT-LIMITED HETEROTROPHIC AND MIXOTROPHIC ENVIRONMENTS[a]

Glucose limitation			Thiosulfate-glucose limitation				
Cell biomass[b]	Residual glucose	Added glucose consumed (%)	Cell biomass[b]	Residual glucose	Residual thiosulfate	Added glucose consumed (%)	Added thiosulfate consumed (%)
39.5	4 μM	~100	76	44 μM	80 μM	~100	~100

- Cells were grown at $D = 0.04$ hr⁻¹ in a New Brunswick Bioflo chemostat. pH was automatically maintained at 6.9 ± 1 by the addition of 5% Na_2CO_3. The concentrations of glucose and thiosulfate in the inflow medium were 0.15 and 1.0%, respectively. To determine the concentration of residual substrates in the medium, steady-state cultures were collected on KCN (final concentration, 5 mM) and rapidly filtered under vacuum. Glucose and thiosulfate concentrations in the filtrates were measured as described.[25] For dry weight determination, washed suspensions of cells were dried to constant weight in vials. Concentrations refer to steady-state values in the chemostat vessel.[55]

- mg (dry weight)/mℓ of culture.

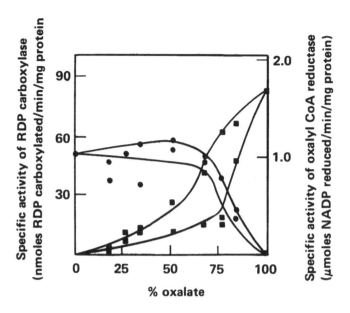

FIGURE 13. Specific activities of Oxalyl CoA reductase (●, ○) and Ribulose diphosphate (RDP) Carboxylase (■, □), in a carbon-limited chemostat culture of *Pseudomonas oxalaticus* at different ratios of formate and oxalate in the feed and two dilution rates (closed symbols, 0.05 hr⁻¹; open symbols, 0.10 hr⁻¹). The percentage of formate in the mixture (total organic substrate concentration 200 mM) is plotted. (From Harder, W. and Dijkhuizen, L., *6th Int. Symp. of Continuous Culture*, Dean A. C. R., Ellwood, D. C., Evans, C. G. T., and Melling, J., Eds., Chichester Ellis-Horwood, Chichester, 1976. With permission.)

has evidently evolved to enable bacteria to adjust efficiently to environments of different nutritional status.

Not only can bacteria shift to different pathways in response to the nutritional status of the environment, they can apparently also preferentially synthesize one form of an enzyme over others. This phenomenon has so far been little studied and only suggestive evidence is available. In a *Spirillum* sp., which seems to be specially adapted to environments of low nutritional status,[16] the pH profile of L-lactate dehydrogenase activity showed three different peaks (Figure 14). During L-lactate limited growth, the relative activites of these enzyme species depended on the culture D value, i.e., the steady-state concentration of L-lactate at which growth occurred. All three activities were higher in cells grown at D = 0.05 hr⁻¹ compared to 0.26 hr⁻¹, but the enzyme species with optima at pH 5.5 and 7.5 to 8 increased approximately fourfold, whereas the species with the optimum pH of 9.5 showed only a twofold increase.[56] It is conceivable that these enzyme species differ in their affinities to L-lactate and other properties, and their selective increase in environments of different nutritional status has adaptive significance. Work is currently in progress to separate and characterize these enzyme species. Changes in the pH profile of phosphatase activity of *A. aerogenes* in response to D have also been reported.[10]

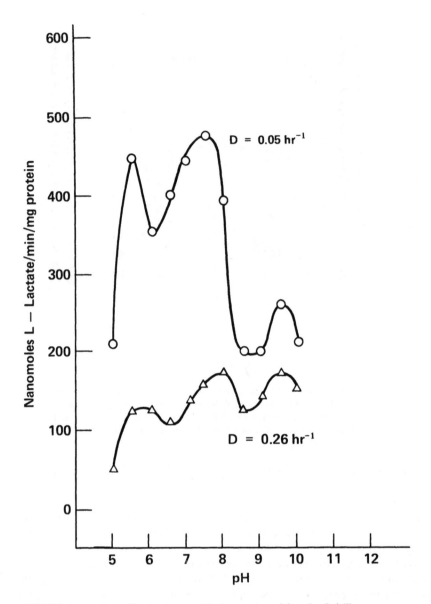

FIGURE 14.   pH profile of L-lactate dehydrogenase activity of a *Spirillum* sp. grown at the specified dilution rates under L-lactate limitation. Unpublished data of A. Matin and J. Steenhuis.

REFERENCES

1. **Duursma, E. K.,** The dissolved organic constituents of seawater, *Chemical Oceanography,* Vol. 1, Riley, J. P. and Skirrow, G., Eds., Academic Press, London, 1965, 433.
1a. **Harvey, R. J.,** Metabolic regulation in glucose-limited chemostat cultures of *Escherichia coli, J. Bacteriol.,* 104, 698, 1970.
2. **Kotze, J. P. and Kistner, A.,** Variations in enzyme activies of *Butyrivibrio fibrisolvens* and Ruminococcus albus grown in continous culture, *Can. J. Microbiol.,* 20, 861, 1974.
3. **Matin, A., Grootjans, A., and Hogenhuis, H.,** Influence of dilution rate on enzymes of intermediary metabolism in two freshwater bacteria grown in continuous culture, *J. Gen. Microbiol.,* 94, 323, 1976.
4. **MacLeod, C. J., Dunnill, P., and Lilly, M. D.,** The synthesis of β-galactosidase by constitutive and other regulatory mutants of *Escherichia coli* in chemostat culture, *J. Gen. Microbiol.,* 89, 221, 1975.
5. **Farago, D. A. and Gibbins, L. N.,** Variation in the activity levels of selected enzymes of *Erwinia amylovora* 595 in response to changes in dissolved oxygen tension and growth rate of D-glucose-limited chemostat cultures, *Can. J. Microbiol.,* 21, 343, 1975.
6. **Silver, R. S. and Mateles, M. I.,** Control of mixed-substrate utilization in continuous culture of *Escherichia coli, J. Bacteriol.,* 97, 535, 1969.
7. **Dean, A. C. R.,** Influence of environment on the control of enzyme synthesis, *J. Appl. Chem. Biotechnol.,* 22, 245, 1972.
8. **Clarke, P. H., Houldsworth, M. A., and Lilly, M. D.,** Catabolite repression and the induction of amidase synthesis by *Pseudomonas aeruginosa* 8602 in continuous culture, *J. Gen. Microbiol.,* 51, 225, 1968.
9. **Hassan, H. M., and Fridovich, I.,** Physiological function of superoxide dismutase in glucose-limited chemostat cultures of *Escherichia coli, J. Bacteriol.,* 130, 805, 1977.
10. **Bolton, P. G. and Dean, A. C. R.,** Phosphate synthesis in *Klebsiella (Aerobacter) aerogenes* growing in continuous culture, *Biochem. J.,* 127, 87, 1972.
11. **Smith, R. W. and Dean, A. C. R.** β-galactosidase synthesis in *Klebsiella aerogenes* growing in continuous culture, *J. Gen. Microbiol.,* 72, 37, 1972.
12. **Hope, G. C. and Dean, A. C. R.,** Pullulanase synthesis in *Klebsiella (Aerobacter) aerogenes* strains growing in continuous culture, *Biochem. J.,* 144, 403, 1974.
13. **Thurston, D. F.,** Induction and catabolite repression of chondroitnase in batch and chemostat cultures of *Proteus vulgaris, J. Gen. Microbiol.,* 80, 515, 1974.
14. **Tempest, D. W., Herbert, D., and Phipps, P. J.,** Studies on the growth of *Aerobacter aerogenes* at low dilution rates in a chemostat, in *Microbial Physiology and Continuous Culture,* Powell, E. O., Evans, C. G. T., Strange, R. E., and Tempest, D. W., Eds., Her Majesty's Stationery Office, London, 1964.
15. **Boddy, A., Clarke, P. H., Houldsworth, M. A. L., and Lilly, M. D.,** Regulation of amidase synthesis by *Pseudomonas aeroginosa* 8602 in continuous culture, *J. Gen. Microbiol.,* 48, 137, 1967.
16. **Matin, A. and Veldkamp, H.,** Physiological basis of the selective advantage of a *Spirillum* sp. in a carbon-limited environment, *J. Gen. Microbiol.,* 105, 187, 1978.
17. **Harder, W., Kuenen, J. G., and Matin, A.,** Microbial selection in continuous culture, *J. Appl. Bacteriol,* 43, 1, 1977.
18. **Botsford, J. L. and Danley, D. E.,** Induction of tryptophanase and accumulation of cyclic AMP in *Escherichia coli* grown in continuous culture, *Abstr. Annu. Meet. Am. Soc. Microbiol.,* 123, 1975.
19. **Rickenberg H. V.,** Cyclic AMP in procaryotes, *Annu. Rev. Microbiol.,* 28, 353, 1974.
20. **Dean, A. C. R.,** Influence of environment on the control of enzyme synthesis, *J. Appl. Chem. Biotechnol.,* 22, 245, 1972.
21. **Smith, R. W. and Dean, A. C. R.,** β-galactosidase synthesis in *Klebsiella aerogenes* growing in continous culture, *J. Gen. Microbiol.,* 72, 37, 1972.
22. **Paigen, K. and Williams, B.,** Catabolite repression and other mechanisms in carbohydrate utilization, *Adv. Microb. Physiol.,* 4, 252, 1970.
23. **DeCrombrugghe, B., Chen, B., Andersen, W. B., Gottesman, M. E., Perlman, R. L., and Pastan, I.,** Role of cyclic adenosine 3', 5' - monophosphate and the cyclic 3', 5' - monophosphate receptor protein in the initiation of lac transcription, *J. Biol. Chem.,* 246, 7343, 1971.
24. **Heinekin, F. G, and O'Connor, R. J.,** Continuous culture studies on the biosynthesis of alkaline protease, neutral protease, and α-amylase by *Bacillus subtilis* NRRL-B3411, *J. Gen. Microbiol.,* 7 35, 1972.
25. **Matin, A. and Rittenberg, S. C.,** Enzymes of carbohydrate metabolism in *Thiobacillus* species, *J. Bacteriol.,* 107, 179, 1971.

26. **Horiuchi, T., Tomizawa, J., and Novick, A.,** Isolation and properties of bacteria capable of high rates of β-galactosidase synthesis, *Biochim. Biophys. Acta*, 55, 152, 1962.
27. **Hartley, B. S.,** Enzyme families, in *Evolution in the Microbial World*, 24th Symp. Soc. Gen. Microbiol., Cambridge University Press, London, 1974, 151.
27a. **Rigby, P. W. J., Burleigh, B. D., Jr., and Hartley, B. S.,** Gene duplication in experimental enzyme evolution, *Nature (London)*, 251, 200, 1974.
28. **Matin, A. and Gottschal, J. C.,** Influence of dilution rate on NAD(P) and NAD(P)H concentrations and ratios in a *Pseudomonas* sp. grown in continuous culture, *J. Gen. Microbiol.*, 94, 333, 1976.
29. **Wimpenny, J. and Firth, A.,** Levels of nicotinamide adenine dinucleotide and reduced nicotinamide adenine dinucleotide phosphate in facultative bacteria and the effect of oxygen, *J. Bacteriol.*, 111, 24, 1972.
30. **Dolezal, J. and Kapralek, F.,** Physiological characteristic of chemostatically grown *Citrobacter freundii* as a function of the specific growth rate and type of nutrient limitation, *Folia Microbio. (Prague)*, 21, 168, 1976.
31. **Chapman, A. G., Fall, L., and Atkinson, D. E.,** Adenylate energy charge in *Escherichia coli* during growth and starvation, *J. Bacteriol.*, 108, 1072, 1971.
32. **Mian, F. A., Fencl, Z., Prokop, A., Mohagheghi, A., and Fazeli, A.,** Effect of growth rate on the glucose metabolism of yeast grown in continuous culture. Radiorespirometric studies, *Folia Microbiol. (Prague)* 19, 191, 1974.
33. **Ng, A. M. L., Smith, J. E., and McIntosh, A. F.,** Influence of dilution rate on enzyme synthesis in *Aspergillus niger* in continuous culture, *J. Gen. Microbiol.*, 81, 425, 1974.
34. **Brown, D. E. and Zainuddin, M. A.,** Growth kinetics and cellulase biosynthesis in the continuous culture of Trichoderma viride, *Biotechnol. Bioeng.*, XIX, 941, 1977.
35. **Fowler, M. W., and Clifton, A.,** Activities of enzymes of carbohydrate metabolism in cells of *Acer pseudoplatanus* L maintained in continuous (chemostat) culture, *Eur. J. Biochem.*, 45, 445, 1974.
36. **Beck, C. and von Meyenberg, H. K.,** Enzyme pattern and aerobic growth of *Saccharomyces cerevisiae* under various degrees of glucose limitation, *J. Bacteriol.*, 96, 479, 1968.
37. **Toda, K.,** Dual control of intertase biosynthesis in chemostat culture, *Biotechnol. Bioeng.*, XVIII, 1117, 1976.
38. **Carter, B. L. A. and Bull, A. T.,** Studies of fungal growth and intermediary carbon metabolism under steady and non-steady state conditions, *Biotechnol. Bioeng.*, XI, 785, 1969.
39. **Huber, T. J., Street, J. R., Bull, A. T., Cook, K. A., and Cain, R. B.,** Aromatic metabolism in fungi. Growth of *Rhodoturula mucilaginosa* in *p*-hydrozybenzoate-limited chemostat and the effect of growth rate on the synthesis of enzymes of 3-oxodipate pathway, *Arch. Microbiol.*, 102, 139, 1975.
40. **Ng, F. M. W. and Dawes, E. A.,** Chemostat studies on the regulation of glucose metabolism in *Pseudomonas aeruginosa* by citrate, *Biochem. J.*, 132, 129, 1973.
41. **Whiting, W. H., Midgley, M., and Dawes, E.,** The regulation of transport of glucose, gluconate and 2-oxogluconate and of glucose catabolism in *Pseudomonas aeruginos, Biochem. J.*, 154, 659, 1976.
42. **Thomas, A. D., Doelle, H. W., Westwood, A. W., and Gordon, G. L.** Effect of oxygen on several enzymes involved in the aerobic and anaerobic utilization of glucose in *Escherichia coli, J. Bacteriol.*, 112, 1099, 1972.
43. **Oura, E.,** Effect of aeration intensity on the biochemical composition of baker's yeast. I. Factors affecting the type of metabolism, *Biotechnol. Bioeng.*, XVI, 1197, 1974.
44. **Oura, E.,** Effect of aeration intensity on the biochemical composition of baker's yeast. II. Activities of the oxidative enzymes, *Biotechnol. Bioeng.* XVI, 1213, 1974.
45. **Wimpenny, J. W. T. and Necklen, D. K.,** The redox environment and microbial physiology. I. The transition from anaerobiosis to aerobiosis in continuous cultures of facultative anaerobes, *Biochim. Biophys. Acta*, 253, 352, 1971.
46. **Rogers, P. J., Clark-Walker, G. D., and Stewart, P. R.,** Effect of oxygen and glucose on energy metabolism and dimorphism of *Mucor genevensis* grown in continuous culture: reversibility of yeast-mycelium conversion, *J. Bacteriol.*, 119, 282, 1974.
47. **Passonneau, J. V., and Lowry, O. H.,** Phosphofructokinase and the Pasteur effect, *Biochem. Biophys. Res. Commun.*, 7, 11, 1962.
48. **Harrison, D. E. F. and Maitra, P. K.,** Control of respiration and metabolism in growing *Klebsiella aerogenes*. The role of adenine nucleotides, *Biochem. J.*, 112, 647, 1969.
49. **Mateles, R. I., Chian, S. K., and Silver, R.,** in *Microbial Physiology and Continuous Culture*, Powell, E. O., Evans, C. G. T., Strange, R. E., and Tempest, D. W., Eds., Her Majesty's Stationery Office, London, 1967, 232.
49a. **Harder, W. and Dijkhuizen, L.,** Mixed substrate utilization in microorgnisms, in *6th International Symposium of Continuous Culture*, Dean, A. C. R., Ellwood, D. C., Evans, C. G. T., and Melling, J., Eds., Chichester: Ellis-Horwood Ltd., Chichester, 1976.

50. Tempest, D. W., Meers, J. L., and Brown, C. M., Synthesis of glutamate in *Aerobacter aerogenes* by a hitherto unknown route, *Biochem, J.*, 117, 405, 1970.
51. Neijssel, O. M., Hueting, S., Carbbendam, K. J., and Tempest, D. W., Dual pathway of glycerol assimilation in *Klebsiella aerogenes* NC1B418, *Arch. Microbiol.*, 104, 83, 1975.
52. Whitting, P. H., Midgley, M., and Dawes, E. A., The role of glucose limitation in the regulation of the transport of glucose, gluconate and 2-oxogluconate, and of glucose metabolism in *Pseudomonas aeruginosa*, *J. Gen. Microbiol.*, 92, 304, 1976.
53. Perez, R. C. and Matin, A., Growth of this *Bacillus novellus* on mixed substrate, *J. Bacteriol.*, 142, 633, 1980.
54. Matin, A., Schleiss, M., and Perez, R. C., Regulation of glucose and metabolism in *Thiobacillus novellus*, *J. Bacteriol.*, 142, 639, 1980.
55. Leefeldt, R. and Matin, A., Growth and physiology of *Thiobacilllus novellus* under nutrient-limited mixotrophic conditions, *J. Bacteriol.*, 142, 645, 1980.
56. Matin, A. and Steenhuis, J., unpublished.

Chapter 4

MICROBIAL BIOENERGETICS DURING CONTINUOUS CULTURE*.**

Walter P. Hempfling and Craig W. Rice

TABLE OF CONTENTS

*   The hitherto unpublished data contained herein were obtained through the support of grants from the National Science Foundation, NO. GB-25582 and No. PCM75-23535.

**   This chapter was submitted in January 1979.

I. INTRODUCTION

The efficiency with which procaroyotes conserve energy liberated during respiration has been a matter of some controversy, with conflicting results arising often during the course of its investigation. Bauchop and Elsden showed that the molar growth yield referred to ATP consumption of a number of anaerobic microorganisms was remarkably constant at about 10.5 g (dry weight) per mole of ATP consumed when carbohydrates served as energy sources.[1] The computations of Gunsalus and Shuster, however, indicated that the observed value was less than one third of that expected based upon the macromolecular composition of the "typical" bacterium, the operation of known pathways of biosynthesis of monomers, and an estimated stoichiometry of ATP consumption during the formation of biopolymers.[13] Forrest and Walker reached similar conclusions nearly a decade later.[11]

Following the appearance of the influential contribution of Bauchop and Elsden, several investigators applied the method of growth yield estimation to the problem of the efficiency of respiration-dependent energy conservation in bacteria, an issue which had defied solution when approached by classical techniques involving the preparation of cell-free extracts. By equating a molar growth yield of 10.5 g/mol of energy source to the production of 1 mol of ATP, it was thought that the stoichiometry of ATP production during substrate metabolism could be determined, and this approach had met with some success in the investigation of the Entner-Doudoroff pathway of glycolysis, which produces only 1 mol of ATP during glucose fermentation as opposed to the production of 2 mol of ATP by the Embden-Meyerhof-Parnas pathway of glycolysis.[13] Upon investigating molar growth yields of *Desulfovibrio* during sulfate respiration Senez, for example, observed values that were too small to permit the conclusion that full coupling of oxidative phosphorylation to respiration was operative.[60] He introduced the term "uncoupled growth" to describe the low-yield phenomenon. According to this view, respiration during low-yield growth might proceed without obligatorily stoichiometric ATP formation, or alternatively, ATP might be synthesized by oxidative phosphorylation and then be partially hydrolyzed by membrane ATPase. The findings by Senez were consistent with observations made with cell-free extracts of bacteria, which only poorly coupled phosphate esterification to oxygen consumption during the oxidation of NAD$^+$-linked electron donors (see, for example, reference 20). Procaryotic oxidative phosphorylation was considered by some investigators to be intrinsically less efficient than that carried out by the mitochondrial respiratory apparatus.

More recently, evidence has become available that demonstrates that aerobic procaryotes are capable of full oxidative phosphorylation, in that P/O values of 3 mol of phosphate esterified per gram atom of oxygen reduced are consistently observed using suspensions of starved, intact *Escherichia coli*.[22] Since the advent of Mitchell's "chemiosmotic" hypothesis,[39] it has become clear that ATP synthesis is only one end product of respiratory energy conservation, and that such energy-requiring reactions as those represented by active transport processes may be considered to be other end products without the intermediate synthesis of high-energy phosphate bonds. Hence, the apparent efficiency of energy conservation may be severely diminished as reflected by molar growth yield estimations, and the discrepancy between predicted and observed yields ought to vary with the identity of the energy source according to the loads imposed upon membrane systems consequent upon the nature of operative transport mechanisms.

It is our purpose to rationalize the dependency of *the molar growth yield upon the* identity of the compound limiting growth by employing explanations drawn from information about the basic biochemical and physiological phenomena of membrane transport processes, respiratory and fermentative ATP synthesis, and of some aspects of macromolecular biosynthesis. We hope to show that the technique of growth yield estimation during continuous culture is an incisive heuristic aid toward enhancing our understanding of microbial bioenergetics.

II. EARLY THEORETICAL CONSIDERATIONS OF THE DETERMINANTS OF POPULATION DENSITY IN CONTINUOUS CULTURE

Under conditions where a continuous culture is truly energy-limited (for example, by oxygen during growth in the presence of an excess of a substrate that will support only respiratory growth, or by a fermentable carbohydrate under anaerobic conditions), we assume, to a first approximation, that the population density, x, is determined by the following expression:[25]

$$x = Y_s (S_R - S_V) \tag{1}$$

where x   = the steady-state concentration of biomass (g/L);
S_R = the concentration of the growth-limiting substrate in the reservoir *(M)*;
S_v = the concentration of growth-limiting substrate in the effluent from the growth vessel *(M)*;
and Y_s = the molar growth yield (g/mol).

We consider further that, under conditions of energy limitation, the magnitude of the molar growth yield is determined by ATP production per mole of substrate (P/S), so that

$$Y_S = Y_{ATP} \cdot P/S \tag{2}$$

This most simplistic set of assumptions results in a pattern of variation of population density with growth rate in substrate-limited continuous culture as predicted by Herbert, et al.[25]

$$x = Y_S \left\{ S_R - K_S \, {}^{(D/\mu_{max} - D)} \right\} \tag{3}$$

where D   = dilution rate, or f/V (f = flow rate of growth medium in L/h and V = culture volume in L);
K_s = substrate concentration allowing half-maximal growth rate;

and μ_{max} = the maximal *extrapolated* specific growth rate. (During steady-state growth in the chemostat the dilution rate, D, is equal to the specific growth rate). As the identity of the substrate is changed, the influence of Y_s upon x will be determined according to the particular value of P/S peculiar to the metabolism of the substrate according to Equation 2.

Significant deviations from the behavior predicted by Equation 3 are observed, however; indeed, so often are disparate results obtained that it is fair to state that Equation 3 describes exceptional variation of x *vs.* D in continuous culture. The dominant departure is one in which lower values of x than would be predicted are found at slow growth rates.

In a seminal communication, Pirt offered a rationale for such a deviation by applying the notion of "maintenance energy" to the behavior of growing bacteria.[49] Maintenance processes are defined as those requiring energy consumption without leading directly to increase of biomass, but which are nevertheless necessary for growth and survival: for example, maintenance of transmembrane disequilibria of solutes and turnover of macromolecular constituents. We will not recount the history of the development of this concept as it has been adequately reviewed elsewhere.[8,33]

Assuming the rate of maintenance metabolism to be independent of the rate of growth of the continuous culture, Pirt derived an expression accounting for the behavior of substrate-limited cultures of microorganisms with significant rates of maintenance metabolism as follows:

$$1/Y_S = m/D + 1/Y_S(\text{max}) \tag{4}$$

where Y_s = the value of the molar growth yield observed at the dilution rate D;

$Y_s(\text{max})$ = the "true" value of Y_s after correction for the diversion of energy for maintenance purposes;

and m = the specific rate of maintenance metabolism of substrate (moles per gram per hour).

Equation 4 allows the determination of the "true" value of the molar growth yield — that is, of Y_s (max) — only by estimating the value of Y_s at several different rates of growth. Pirt calculated, for example, that the rate of glucose consumption for purposes of maintenance during continuous culture of *Aerobacter cloacae*[49] was 0.5 mmol per g-h during aerobic growth and 2.6 mmol per g-h during anaerobic growth; values of $Y_{glucose}$ (max) were 79 g/mol and 15 g/mol, respectively.

Pirt's treatment of yield as affected by growth rate led to modification of Equation 3 by substituting Equation 4 in the Y_s term to result in the following expression:

$$x = DY_S(\text{max})/mY_S(\text{max}) + D \left\{ D_R - K_S (D/\mu_{max} - D) \right\} \tag{5}$$

III. PRACTICAL CONSIDERATIONS

An array of data describing the aerobic and anaerobic yield and maintenance characteristics of *Escherichia coli* B obtained in the authors' laboratory is presented in Table 1. These values were calculated by the application of Equation 4 to molar growth yield data gathered at numerous different growth rates. A variety of carbon sources was employed to limit growth under aerobic and anaerobic conditions. For comparative purposes, several other nutrients were used to limit growth with succinate or mannose as carbon source. The same minimal medium with ammonium ion used as nitrogen source was employed throughout.

In further experiments, 8 mM acetate or formate added to aerobic continuous cultures otherwise limited by mannose resulted in diminution of the values of Y_o (max) to 21.4 and 18.4 gram per gram atom oxygen, respectively. Acetate (which was completely oxidized) raised the rate of maintenance respiration to 7.8 mg atoms oxygen per gram per hour, but the value of m_o of the formate-supplemented culture was essentially unchanged at 2.0 mg atoms 0 per gram per hour. Formate was not oxidized in the presence of mannose.

Although changing the identity of the carbon source limiting growth resulted in var-

Table 1
GROWTH YIELD PARAMETERS OF *ESCHERICHIA COLI* B
IN CONTINUOUS CULTURE

Molar Growth Yields and Maintenance Requirements Referred to ATP in Anaerobic
Continuous Culture

Growth-limiting substance	Carbon source	Values[a] of μ	Y_{ATP} (max)[b,c]	m_{ATP}[b,d]	Ref.
Mannose	Mannose	7	11.7 (10.8—12.8)	6.3 ± 0.08	55
Fructose	Fructose	5	11.0 (10.6—11.3)	11.2 ± 0.3	55
Sorbitol	Sorbitol	6	11.3 (10.2—12.7)	7.6 ± 1.1	55
Mannitol	Mannitol	7	12.5 (11.0—14.4)	7.8 ± 1.0	55
Glucose	Glucose	6	10.7 (9.5—12.1)	20.0 ± 1.4	24
Galactose	Galactose	7	9.3 (8.8— 9.7)	9.3 ± 0.7	55
Sulfate	Mannose	7	9.4 (8.9—10.0)	13.0 ± 1.2	55

Molar Growth Yields and Maintenance Requirements Referred to Oxygen Consumption in Aerobic Continuous Culture

			Y_O(max)[b,e]	m_O[b,f]	
Glutamate	Glutamate	10	34.7 (28.8—43.7)	20.4 ± 1.0	24
Gluconate	Gluconate	8	28.1 (27.1—29.3)	1.7 ± 0.2	55
Acetate	Acetate	4	14.7 (13.5—16.1)	25.4 ± 1.3	24
Mannose	Mannose	8	31.6 (29.9—33.6)	2.6 ± 0.2	55
Phosphate	Mannose	9	34.2 (28.6—42.2)	15.6 ± 0.6	55
Sulfate	Mannose	6	26.7 (22.2—33.4)	8.7 ± 0.7	55
Succinate	Succinate	6	11.1 (10.0—12.4)	11.7 ± 1.2	24
Oxygen	Succinate	7	12.5 (12.0—13.0)	3.3 ± 0.4	54
Phosphate	Succinate	7	12.6 (11.6—13.8)	6.1 ± 0.9	55
Sulfate	Succinate	8	14.4 (12.8—16.4)	6.7 ± 0.9	55

[a] Number of different growth rates examined.
[b] Value and 95% confidence limits by linear regression analysis of best fits by least squares.
[c] Gram (dry weight) per mole ATP.
[d] Millimoles ATP per gram per hour.
[e] Gram (dry weight) per gram atom oxygen consumed.
[f] Milligram atoms oxygen per gram per hour.

iation of the value of Y_o (max) from about 11 gram per gram atom O to nearly 35 gram per gram atom O, limiting the growth of cells supplied with succinate by restricting the amounts of sulfate, phosphate, or oxygen produced little effect upon the values of Y_o (max) of those continuous cultures. Limitation of mannose-supplied cultures with sulfate brought about a significant diminution of the value of Y_o (max).

The measurement of the values of Y_{ATP} (max) as previously described[24] disclosed that the parameter varied little with the identity of the carbon source limiting anaerobic growth so long as the carbohydrate was transported by the PEP:sugar phosphotransferase system.[51] Changing the identity of the substance limiting anaerobic growth to galactose (not a so-called "PTS sugar") or to cultures growing with excess mannose, but limited by sulfate, brought about appreciable changes in the observed values of Y_{ATP} (max).

Through combination of the data derived from yield studies of aerobic and anaerobic continuous cultures, we hope to achieve reliable estimates of the efficiency of oxidative phosphorylation in growing *E. coli*.[24] By applying Equation 2 to some of the data contained in Table 1, we have suggested that yields lower than those expected are due to variable coupling between respiration and phosphate esterification, that is, variation in the value of P/S occurs consequent upon growth at the expense of different energy sources.[19,21,24] Others have added the notion of "growth rate-dependent" maintenance metabolism,[45] the magnitude of which is dependent upon the identity of the substrate limiting growth, as well as upon the rate of growth. Although the biosynthetic demands upon the energy supply are greater in the cases of growth limited by succinate or acetate than they are during growth limited by a carbohydrate,[62] the magnitude of such added biosynthetic requirements are insufficient to account for the threefold differences in P/O values noted.

We believe that it will be most helpful to modify the Pirt expression given in Equation 4 to include variation in the energetic expense of monomer biosynthesis and of energy required for transport of necessary compounds into the growing cell. We prefer to resort to the notion that variation in the efficiency of oxidative phosphorylation is responsible for the observed variation in aerobic molar growth yields only if more conservative explanations embodied in our modification of Equation 4 prove insufficient to account for experimental results.

IV. A DETERMINISTIC MODEL RATIONALIZING THE SUBSTRATE DEPENDENCY OF MOLAR GROWTH YIELD

The rate of overall substrate utilization necessary to supply energy for growth rate-independent maintenance processes, net transport processes, conversion of substrate into monomers, and polymerization of those monomers into biopolymers is given by

$$-dS/dt = (dS/dt)_M + (dS/dt)_T + (dS/dt)_C + (ds/dt)_P \qquad (6)$$

where $(dS/dt)_M$ = the rate of substrate utilization necessary to supply energy to growth rate-independent maintenance processes;

$(dS/dt)_T$ = the rate of substrate utilization necessary to supply energy to growth rate-dependent transport processes;

$(dS/dt)_C$ = the rate of substrate utilization necessary to supply energy for the conversion of substrate into monomers;

and $(dS/dt)_P$ = the rate of substrate utilization necessary to supply energy for the polymerization of monomers into biopolymers.

By making the following substitutions

$(dS/dt)_M$ = $-mx$;

$(dS/dt)_T$ = $-\mu x r$;

$(dS/dt)_C$ = $-\mu x \varrho$;

and $(dS/dt)_P$ = $-\mu x/Y_P$, Equation 6 becomes

$$1/Y_S = m/\mu + r + \rho + 1/Y_P \qquad (7)$$

The value of the term $1/Y_s(max)$ in Equation 4 (Pirt's original expression), by definition, is given by the sum of all growth rate-dependent processes, or

$$1/Y_S(max) = r + \rho + 1/Y_P \qquad (8)$$

The amount of substrate consumed to provide the energy necessary for *net* transport of all materials required for the unit increase of biomass, including the sources of carbon, nitrogen, phosphate, and sulfur is given by the term r in Equation 8. As the carbon source is the source of energy in heterotrophs, r must include an allowance for the energy expended to transport sufficient energy source to drive *all* energy-requiring reactions that occur at rates proportional to the rate of growth. The term ϱ represents the amount of substrate consumed to supply all energy necessary for the conversion of the sources of carbon, nitrogen, phosphate, and sulfur into the monomers used for the synthesis of biological polymers in a unit of biomass, and will clearly vary with the identity of those elemental sources. The term $1/Y_P$ represents the amount of substrate required to furnish energy for the polymerization of monomers (amino acids, ribo- and deoxyribonucleotides, monosaccharides, and fatty acids) to form the biopolymers of a unit of biomass. There is no reason to expect that the value of $1/Y_P$ will vary as the identity of the substance limiting growth in continuous culture is changed. According to our estimation, the change of macromolecular composition of a bacterial population attending the change of the rate of growth of that population is small enough to have a negligible effect upon the value of $1/Y_P$. Significant variation of any of the component terms in Equation 8 as the rate of growth is changed would be indicated by nonlinearity of a plot of $1/Y_s$ vs. $1/D$.

Although the parameters r and ϱ comprise components that account for the transport and intermediary biosynthetic components of all cellular constituents, the source of carbon and energy and the source of nitrogen dominate those terms, because those compounds supply the predominant elemental components of the bacterial cell. The magnitude of the value of r depends upon the mechanism of transport of each elemental source. For example, when a hexose transported by the sugar:PEP phosphotransferase system (PTS) serves as carbon source the value of r ought to be smaller than when a compound whose net transport brings about net energy expenditure supplies carbon. The value of ϱ will similarly be affected by the identities of the sources of carbon and nitrogen: for example, increasing values of ϱ would be expected as the carbon source is changed from glucose to lactate to acetate, since monomer synthesis becomes increasingly energetically expensive.

The values of r and ϱ of the cultures described in Table 1 are presented in Table 2.

Equation 8 permits the calculation of the values of $Y_{ATP}(P)$ and $Y_O(P)$ from the data compiled in Tables 1 and 2. The amounts of biomass formed by the application of energy solely to polymerization reactions are given in Table 3. The calculation of the value of $Y_O(P)$ for each culture requires, in addition to the aforementioned assumptions, an estimate of the efficiency of oxidative phosphorylation. In order to make this calculation we have assumed either a P/O value of 2 or 3 mol P_i esterified per gram atom of oxygen consumed (Table 3). For example, the value of $Y_O(P)$ of mannose-limited cultures is given by the following calculation:

$$1/Y_O(P) = 1/Y_O(max) - (r \times P/O^{-1} + \rho \times P/O^{-1}) \qquad (9)$$

Substituting the values from Table 2 for r and ϱ, and assuming a maximal P/O value of 3 (2.9 in the case of mannose-grown cells) we have

$$1/Y_O(P) = 0.03597 - (0.014/2.9 + 0.0066/2.9) \qquad (10)$$

and the value of $1/Y_O(P)$ is therefore 0.0289 g atom oxygen per gram. Similarly, an assumption of a maximal P/O value of 2 leads to the following expression,

$$1/Y_O(P) = 0.03597 - (0.014/1.83 + 0.0066/1.83) \qquad (11)$$

Table 2
CALCULATION OF R AND ϱ FOR *ESCHERICHIA COLI* B IN CONTINUOUS CULTURE

Carbon source	Substrate[a] incorporated (mmol/g)	Substrate[b] catabolized (mmol/g)	ATP/ substrate[c] transported (mol/mole)	mg-atom or[c] mg-ion/g	r (mmoles ATP/g)	ϱ[d] (mmoles ATP/g)
Anaerobic Growth[p]						
"PTS"[o]	6.3	29.1	0[f]	14.0	14	6.6
Sugars						
Galactose	6.3	35.5	0.5	14.0	34.9	6.6
Aerobic Growth[e]						
Mannose	6.3	3.0	0[f]	14.0	14	6.6
Gluconate	6.3	4.2	1	14.0	24.5	6.6
Glutamate	7.6	4.4	1	3.0[g]	15	10.9[h]
Acetate	19.0	17.0	1	14.0	50	50.7
Mannose *plus*						
Acetate[i]	28.5[j]	11.7[k]	1[m]	14.0	29	9.5[n]
Succinate	9.5	14.2	1	14.0	37.7	25.6

[a]　Substrate incorporated = 0.456 g atom C/g *times* moles substrate/g atom C (Ref. 24)

[b]　Substrate catabolized = $[Y_o$(max), corrected for substrate level phosphorylation]$^{-1}$ *times* moles substrate completely oxidized per pair of reducing equivalents produced, *plus* the amount of substrate oxidized to offset the difference between 6.3 mmol hexose per gram and the amount of substrate incorporated.[a] Substrate catabolized in anaerobically grown cells = $[Y_{ATP}$ (max)]$^{-1}$ *times* (moles of substrate catabolized per mole ATP generated, which we assume to reach a maximum of three according to the following reactions: (i) glucose + 2 ADP + 2 P_i → 2 pyruvate + 2 ATP + 2 (2e⁻ + 2H⁺); (ii) pyruvate + ADP + P_i → acetate + formate + ATP).

[c]　The sum of the gram-atom or gram-ion content per gram (dry weight) of N (12.25 mg atom per gram), K (0.7 mg ion per gram), P (0.7 mg atom + mg ion per gram), and S (0.35 mg atom per gram). In the calculation of the value of r a stoichiometry of one phosphate anhydride bond hydrolyzed per NH_4^+, K^+, PO_4^{3-} or SO_4^{2-} ion is assumed (also see text).

[d]　ϱ = the calculated ATP requirement for the conversion of substrate incorporated[a] into 6.3 mmol hexose per gram *plus* the energy expended for the synthesis of all monomers per gram cell mass from hexose, NH_4^+, SO_4^{2-}, and PO_4^{3-} based on known pathways (Ref. 32) and the equivalent monomeric composition of the cell (Ref. 11).

[e]　Cells grown aerobically at 37°.

[f]　With group translocation[51] no net phosphate anhydride bond hydrolysis is required for transport.

[g]　Corrected for 90% of cellular nitrogen derived from the amino group of glutamate.[35]

[h]　Calculated using the most direct pathways for the conversion of glutamate into cell monomers.

[i]　Supplied in the growth medium in the ratio of 1 mol mannose to 4 mol acetate (see text).

[j]　Based on the experimental determination that about 15% of cellular carbon arises from acetate in these cultures. Mannose incorporation not included since its transport does not affect the calculation of the value of r(see [f] above)

[k]　Assuming that all oxygen is reduced by reducing equivalents arising from acetate.

[m]　For acetate transport only.

[n]　Assuming that all acetate incorporated enters fatty acids with a stoichiometry of 1 acetate per ATP.

[o]　Sugars transported by the phosphoenolpyruvate: sugar phosphotransferase system.[51] Y_{ATP} (max) = 11.44 g/mol ATP, the mean value for cultures limited by glucose, fructose, mannitol, sorbitol, and mannose (see Table 1).

[p]　Cells grown anaerobically at 37°.

[q]　See text.

Table 3
CALCULATION OF Y_{ATP} (P) and $Y_o(P)$ OF *ESCHERICHIA COLI* B IN CONTINUOUS CULTURE

Anaerobic growth

Carbon[*] source	Y_{ATP}(max)	Y_{ATP}(P)[b]
"PTS-sugars"	11.4	15.0
Galactose	9.3	15.3

Aerobic growth

Carbon[c] source	Y_o (max)[d]	Apparent[e] P/O	Assuming P/O = 3			Assuming P/O = 2		
			$Y_o(P)$[f]	Predicted[g] P/O	Calculated[h] P/O	$Y_o(P)$[f]	Predicted[g] P/O	Calculated[h] P/O
Mannose	27.8	2.43	34.65	2.9	2.31	40.5	1.83	2.70
Gluconate	24.9	2.18	34.49	2.82	2.30	43.6	1.82	2.90
Glutamate	30.9	2.70	43.4	2.8	2.89	53.6	1.89	3.58
Acetate	14.7	1.28	31.84	2.75	2.12	95.4	1.75	6.36
Mannose *plus* Acetate	21.4	1.87	30.6	2.75	2.03	40.4	1.75	2.70
Succinate	11.1	0.97	15.0	2.70	1.0	19.2	1.66	1.28

[*] Limited by carbon source (anaerobic).
[b] Calculated directly from Equation 8 and data in Table 2.
[c] Limited by carbon source (aerobic).
[d] Corrected for substrate level phosphorylation.
[e] Y_o (max)[d]/Y_{ATP} (max). Values from Table 1.
[f] Calculated from Equation 8 and data in Table 2 divided by the appropriate P/O value (see Equation 9).
[g] Calculated by assuming complete oxidation of substrate and three "sites" of oxidative phosphorylation associated with pyridine nucleotide oxidation and two "sites" of oxidative phosphorylation associated with flavoprotein oxidation in cells capable of P/O = 3 and two and one "sites" of oxidative phosphorylation, respectively, cells capable of P/O = 2.
[h] $Y_o(P)$ /$Y_{ATP}(P)$.

giving a value of $1/Y_o$ (P) of 0.0247 gram atom oxygen per gram. Since the values of Y_o (max) given in Table 3 and used for these calculations have been corrected for the contribution to biomass formation of substrate-level phosphorylation, the efficiency of oxidative phosphorylation may be estimated by Equation 2; accordingly:

$$P/O = Y_O(P) / Y_{ATP}(P) \qquad (12)$$

where the value of Y_{ATP} (P) is that obtained from anaerobic continuous culture and given in Table 3.

As can be seen from inspection of Table 3, an assumption of maximal oxidative phosphorylation efficiency of 3 mol P_i esterified per gram atom of oxygen consumed leads to better agreement between predicted and observed P/O values than does a P/O value of 2, although some residual discrepancy remains, especially in the case of cultures supplied with succinate. The assumption of an operational P/O value near 3 is buttressed, however, by direct P/O measurements using starved, intact bacteria.[22]

To the extent that the assumptions required in these calculations are unrealistic, the calculated values of Y_{ATP} (P) and Y_o(P) will, of course, be in error. Such inaccuracies should not be construed as a fundamental failing of the model, however, as assumptions are necessary in the absence of appropriate data.

Uncertainties in the calculation of values of r impose the most likely sources of error. We assume that the source of carbon is either incorporated totally or oxidized in part to CO_2. The actual production of other metabolic end products will lead to an underestimation of the amount of substrate catabolized in proportion to the extent of their formation. A second source of possible error is the assignment of a given stoichiometry of phosphate anhydride bonds hydrolyzed per mole of substrate transported in a net manner. It is tempting to accept the predictions of the chemiosmotic view of solute transport by assuming that anions are taken up coupled to sufficient protons so as to achieve charge neutrality, and that cations are similarly transported in exchange for compensatory proton extrusion. As pointed out by Harold,[15] however, the transmembrane concentration gradients of solutes such as gluconate, glucose-6-phosphate, succinate, and methylamine (an analog of NH_4^+ ion — see Ref. 61) are often larger than can be rationalized by charge-mediated transfer and transmembrane pH differences of realistic magnitude. Inorganic phosphate (P_i) is taken up with concomitant proton consumption, but ATP hydrolysis is also required for its transport.[15] For the purposes of transport of sulfate[4] and potassium ions[53] energy is also required. Based upon these observations we have assumed a minimal value of 1 mol of ATP hydrolyzed to ADP and P_i per mole of solute transported in order to perform the calculations given in Table 2. Such an assumption may not be valid for acetate transport, as that compound equilibrates across the membrane in membrane vesicles in direct response to the pH gradient imposed.[52] Nevertheless, the general absence of ATP-driven transport systems in vesicles and the apparent existence of *two* transport systems for acetate in intact cells[58] makes the operation of ATP-dependent acetate transport in the growing cell a reasonable assumption. A requirement of 0.5 mol ATP per mole of galactose transported has also been calculated.[66]

To the degree that our knowledge of intermediary metabolism and of the biosynthesis of biological monomers is deficient, the calculation of the value of ϱ for each substrate will be in error.

Based upon the foregoing considerations we draw the following conclusions about the utility of Equation 8 toward rationalizing the substrate-dependent differences in molar growth yields:

1. The difference between the values of Y_{ATP} (max) manifested by anaerobic cultures limited by PTS sugars on the one hand and by galactose on the other can be rationalized by the growth rate-dependent difference between the energy required for transport of the sugars (difference in the values of r);

2. The differences among the values of Y_o(max) of cultures growing under limitation by mannose, gluconate, glutamate, or acetate can be reasonably accounted for by differences among the growth rate-dependent expenditure of energy for net transport of the carbon and energy source and the varying expense of monomer synthesis when it is assumed that the efficiency of oxidative phosphorylation is close to 3 mol of inorganic phosphate esterified per gram atom of oxygen consumed.

Having demonstrated that Equation 8 and the operation of fully efficient oxidative phosphorylation can account for differences among growth yields manifested by some cultures, we now wish to consider why such a satisfactory conclusion has not been reached with cultures growing at the expense of succinate, mannose *plus* acetate, and mannose *plus* formate, which all exhibit, to varying degrees, molar growth yields lower than can be accounted for by the hypothesis. We will refer to this phenomenon as "yield shortfall".

V. YIELD SHORTFALL

We may immediately dismiss the example of the culture growing at the expense of mannose *plus* formate, because Equation 8 cannot be applied in the absence of information about the rate of *net* transport of formate into the cell, and therefore cannot be tested. It is clear, however, that through some growth rate-dependent process the monocarboxylic acid is capable of lowering the value of Y_o (max) without similarly affecting the rate of maintenance respiration, and that this occurs without oxidation of the acid.

The two remaining instances of yield shortfall are sufficiently tractable to prompt a discussion of possible reasons for their occurrence. We will consider the following possible mechanisms: "uncoupling" of phosphorylation from respiration in otherwise competent cells by so-called "slip reactions"; "uncoupling" of phosphorylation from respiration by possible repression of synthesis of a necessary component of the apparatus of oxidative phosphorylation; and bypass of phosphorylating electron transport by alternate electron transfer mechanisms.

A. "Slip Reactions"

The terms "uncoupled growth,"[60] "overflow metabolism",[44] "slip reaction,"[46] and "energy spilling"[46] have all been used to describe the apparent inefficiency of utilization of ATP for growth. This concept of "wanton waste" is usually thought to operate through the activity of membrane ATPase, which hydrolyzes ATP that would otherwise be available to drive growth-linked processes. Neijssel and Tempest[44] distinguish between "carbon-limited" and "carbon-sufficient" growth conditions, claiming that the former condition results in energy dissipation by "slip reactions". According to their interpretation, the difference between the value of Y_o(max) of mannose- and acetate-limited cultures would be, in large part, due to the unavailability of sufficient carbon skeletons during acetate-limited growth to permit full application of conserved energy to biomass formation. The major consequence is loss of energy and hence inefficient growth with acetate as carbon and energy source when compared to mannose-limited growth.

We believe that the experiments upon which this view of yield shortfall is based have been improperly designed and poorly interpreted. We wish to consider those experiments in some detail.

When the component limiting growth of *Klebsiella aerogenes* NCTC 418 in continuous culture was sulfate, phosphate, or ammonia, Neijssel and Tempest found that the aerobic growth yield measured at a specific growth rate of 0.17 hr^{-1} was markedly diminished as compared to the yield observed when the carbon source (glucose, glycerol, mannitol, or lactate) served to limit growth.[44] Accompanying this diminution of apparent molar growth yield was the production of appreciable amounts of acetate, regardless of the identity of the carbon source, as well as *alpha*-ketoglutarate during ammonium limitation, along with several other organic acids. Limitation of growth by sulfate, ammonium, or phosphate in the presence of glucose led to the formation of large amounts of polysaccharide — which accounted for from 7 to 33% of the glucose utilized. The estimations of molar growth yield could not be properly interpreted, since the effect of growth rate upon yield was not assessed. Ignoring this consideration, however, the authors explained the diminished values of Y_o as only another aspect of "overflow metabolism"; one in which energy otherwise conserved was dissipated through ". . . either ATP-spilling reactions . . .", ". . . or else that the transfer of electrons to oxygen can occur without concomitant oxidative phosphorylation".

For some time it has been accepted that the mechanism of respiratory control dependent upon the availability of inorganic phosphate acceptors, as found in isolated mitochondria, apparently does not operate in non-growing bacteria.[23] Nevertheless, this observation does not necessarily signify the absence of respiratory control in growing bacteria, as other mechanisms may well be operative. Neijssel and Tempest went on to examine the respiration of *K. aerogenes* following cessation of medium flow in an otherwise phosphate-limited culture in the presence of excess glycerol,[45] and found that as the growth rate prior to stagnation was increased above about 0.3 hr^{-1} the amount of oxygen consumption persisting after the "step-down" also increased. Below a specific growth rate of about 0.2 hr^{-1} a basal respiratory rate of about 7 mmol O$_2$ per gram per hour was manifested, and this rate did not change over about 24 min of observation. It was assumed that growth ceased at the moment of halting the flow of medium. The authors interpreted these results as indicating a progressive loss of respiratory control as the growth rate was decreased. It must be pointed out, however, that the growth rate of bacteria in nutrient-limited continuous culture is determined by the steady-state concentration of that nutrient, which is in turn set by the interplay among metabolism, medium addition, and culture overflow. The cessation of medium flow does not in fact immediately terminate cellular growth, but instead results in a gradual decline of growth rate dictated by the instantaneous concentration of the progressively disappearing limiting nutrient. During phosphate-limited continuous culture, faster growth rates are obtained because more phosphate is available than at slower growth rates, and it is to be expected that respiration *and growth* will persist longer in such cultures than in more slowly growing ones. Furthermore, analysis of the data by means of Equation 4 discloses that at the slowest growth rate examined (0.08 hr^{-1}) some 85% of the respiration is devoted to purposes of maintenance and should not be much affected by changing the growth rate downwards; at the most rapid growth rate (0.62 hr^{-1}) only about 40% of the respiration serves maintenance processes, so that termination of medium flow ought to lead to a diminution of respiration of about 60%. The conclusions drawn by Neijssel and Tempest from this experiment are not supported by the data.

In a further communication,[46] Neijssel and Tempest reported that the addition of saturating amounts of glucose to glucose-limited continuous cultures and of phosphate to phosphate-limited continuous cultures of *K. aerogenes* results in increase of the rate of respiration, the increment being greatest at the slowest growth rate and smallest at the most rapid growth rates. Again, the population density changes following perturbation of the steady states of growth were not reported, so that it is not possible to evaluate their allegation that " . . . energy-spilling reactions must be evoked immediately". Nevertheless, unlike the previous experiment in which the growth rate was stepped down, it seems unlikely that the bacteria would be able to attain their maximum growth rate within a minute after nutrient saturation if for no other reason than a limitation of protein synthesis due to an insufficiency of ribosome content.[29] As this experiment is analogous to the addition of glucose to non-growing cells, it seems intuitively obvious that respiration must be in part uncoupled from growth. It therefore adds no new information to our present store.

In the same publication, Neijssel and Tempest compare growth of a glucose-limited aerobic continuous culture of *K. aerogenes* in the presence and absence of 0.34 *M* NaCl. They assume that a transmembrane concentration gradient is imposed upon the cells as a consequence of the presence of the salt, although evidence for the existence of such a gradient is not presented. The dependency of the specific respiratory rate upon the specific growth rate and the rate of maintenance respiration were unaffected by the salt, observations that were interpreted by the investigators as an indication of the availability of otherwise dissipated excess energy in this so-called "carbon-limited" culture, which is applied to maintain the assumed sodium ion gradient. The essential control experiment — namely, an examination of the effect of NaCl addition upon yields of continuous cultures under energy limitation — was not performed. Rather than demonstrating " . . . that glucose-limited cultures were not energy-limited . . . " we believe that a maintenance load was imposed upon the cells insufficient to have an appreciable effect upon the growth yield. Even at a concentration of 1 M NaCl in the growth medium, Watson was able to demonstrate an intracellular/extracellular sodium gradient of only 10:1 during glucose-limited continuous culture of *Saccharomyces cerevisiae*. This was accompanied by an increment of maintenance expenditure of ATP of only 1.7 mmol per gram per hour.[68]

As shown in Table 1 limitation of the growth of *Escherichia coli* B by mannose or by sulfate in the presence of excess mannose, as well as by succinate or by phosphate, sulfate, or oxygen in the presence of excess succinate, results in values of Y_o (max) that are characteristic of the identity of the carbon source. If the phenomenon of "energy spillage" were operative in these cultures one would expect to find a dependency of the value of Y_o(max) upon the nature of limitation of the continuous culture. In those few cases where the value of Y_o (max) is affected by the limiting nutrient, other explanations are available for the change of growth yield (see below).

There is no compelling reason to invoke hypothetical "slip reactions" used as a safety valve to allow excess energy to escape. The balance of phosporylation and dephosphorylation is dictated by the peculiarities of growth on specific substrates. We prefer not to distinguish between "carbon-sufficient" and "carbon-insufficient" growth conditions, as such a concept is founded upon an unsupported notion that catabolism of the carbon source and related energy conservation are normally dissociated phenomena.

B. Yield Shortfall through Alteration of the Apparatus of Oxidative Phosphorylation

Three cases of yield shortfall through possible repressive effects of growth conditions upon the apparatus of oxidative phosphorylation have been described. These are repression of oxidative phosphorylation during growth on certain carbohydrates, diminution of the efficiency of respiration-dependent proton extrusion during growth with limiting oxygen, and a decrease of respiration-dependent proton extrusion during growth limited by sulfate.

1. Using an early method of assay of oxidative phosphorylation in intact, nondividing cells of *Escherichia coli* B[19] Hempfling claimed that growth in the presence of some carbohydrates lowered the P/O value below that observed in cells that had been grown in a medium containing a mixture of amino acids as carbon sources. The effect of growth in the presence of glucose was particularly pronounced (P/O value less than 1 mol P_i esterified per gram atom of O), and addition of 3′, 5′-cyclic adenosine monophosphate (cAMP) was found to restore phosphorylation efficiency to about 3 mol P_i esterified per gram atom O.[21] Measurement of the aerobic molar growth yield at several growth rates during glucose-limited continuous culture disclosed that the value of Y_o (max) was 12.5 gram per gram atom O consumed, and that the addition of cAMP to the continuous culture brought the aerobic molar growth yield corrected for increased maintenance respiration to about 31 gram per gram atom O.[24] It was suggested that catabolite repression of synthesis of some component of the apparatus of oxidative phosphorylation was responsible for the effect of glucose, and that cAMP reversed the consequences of glucose-dependent growth in a manner similar to the mechanism

of the effects of those compounds on expression of the *lac* operon. Other workers had remarked upon the apparently poor efficiency of oxidative phosphorylation performed by particulate subcellular fractions of glucose-grown *E. coli*,[28] as well as upon the diminished stoichiometry of respiration-dependent proton uptake by inverted vesicles prepared from glucose-grown *E. coli*.[26]

We now have reason to reject the earlier interpretation of our finding concerning the glucose effect upon oxidative phosphorylation, as the ability of our isoalte of *E. coli* B to respond to the presence of glucose by growing inefficiently is no longer displayed. Upon comparing the properties of responsive and nonresponsive cultures we discovered that, unlike the responsive cultures (which produced as much as 0.7 mol of acetate per mole of glucose oxidized during aerobic growth) our present, unresponsive cultures produced no acetate during aerobic glucose-limited growth. Measurement of the efficiency of oxidative phosphorylation in intact nonresponsive cells by a much-improved assay procedure[22] showed that, although the P/O value of mannose-grown cells was close to 3 mol P_i esterified per gram atom of oxygen consumed, the P/O value fell to about 1 when 1 mM formate or acetate was added to the anaerobic cell suspension prior to oxygenation. Moreover, during the estimation of respiration-dependent proton extrusion, the inclusion of less than 100 μM acetate in the anaerobic cell suspension prior to oxygenation resulted in an enhanced rate of reabsorption of protons when the time course of pH change was compared to that of the same starved cell suspension assayed in the absence of added carboxylic acid.[55] Finally, when included in the growth medium during otherwise mannose-limited continuous culture, formate or acetate lowered the value of Y_O (max) (described above).

We accordingly attribute our earlier results showing an apparent uncoupling effect of growth at the expense of some carbohydrates to the presence of metabolically produced acetate in the culture fluid. The major difference between the earlier "responsive" cultures and the later "nonresponsive" cultures is then the inability of the former to oxidize acetate during growth in the presence of even small amounts of glucose, leading to accumulation of the organic acid, while the latter are capable of acetate oxidation during growth at the expense of carbohydrates. The earlier continuous culture experiments in which yield shortfall was noted during growth limited by glucose is analogous to the present experiments reported in Table 1, in which yield shortfall is elicited by growth in the presence of formate limited by mannose, as formate was not oxidized. Moreover, as shown by the effect of added acetate in Table 1, even the ability to oxidize acetate does not protect against yield shortfall if the acetate is exogenously supplied. The addition of cAMP to glucose-containing cultures of "responsive" cells would be expected to bring about derepression of synthesis of enzymes necessary for complete oxidation of acetate formed endogenously, thereby eliminating yield shortfall through restoration of fully efficient oxidative phosphorylation. Inadequate starvation of cells from acetate-containing cultures is the probable reason for the observation of lowered P/O values obtained during measurement of oxidative phosphorylation.[19]

What we had interpreted as a primary effect of glucose metabolism upon the synthesis of components of the apparatus of oxidative phosphorylation was, therefore, according to the present hypothesis, only a secondary effect of the metabolic accumulation of extracellular acetate. We conclude that the phenomenon of catabolite repression of oxidative phosphorylation is not operative.

2. Moss showed more than 25 years ago that the complement of terminal cytochrome oxidases of *E. coli* was subject to modification in response to the availability of oxygen during growth, in that cytochrome a_2 *(d)* was synthesized in significant amounts only during continuous culture under conditions of restricted oxygen supply.[42] *Using this phenomenon as a means to vary the amounts and identities of terminal oxidases, Meyer*

and Jones reported that, along with changes of terminal oxidases, limitation of the growth rate of *E. coli* through varying the rate of oxygen supply led to alteration of the efficiency of respiration-dependent proton extrusion.[37] Bacteria containing cytochrome *d* in addition to cytochrome *o* were found to grow less efficiently and to manifest lower H^+/O values than bacteria respiring with cytochrome *o* as sole terminal oxidase. Accordingly, it was suggested that oxidative phosphorylation functioned less efficiently during respiration terminated by cytochrome *d* than during terminal electron transfer terminated by cytochrome *o*.

That interpretation of results was criticized by Rice and Hempfling, who also provided a critique of methods used to cultivate bacteria continuously under conditions of oxygen limitation.[54] Rice and Hempfling cultivated *E. coli* B continuously under oxygen limitation with excess succinate in the phauxostat, a device in which the rate of growth medium addition from the reservoir is controlled by the growth-dependent change of pH in the culture vessel.[34] The rate of change of the pH of the growth medium was set by the rate of aeration and the rate of respiration, which in turn could be adjusted by varying the steady-state population density through manipulation of the buffering capacity of the inflowing medium in the reservoir. The results demonstrated that neither the value of Y_o (max) nor the value of H^+/O was dependent upon the identities of terminal oxidases present in cells grown with either succinate or oxygen limitation (see also Table 1). The only yield parameter to change significantly was the rate of maintenance respiration (Table 1). Direct measurements of the P/O value of *E. coli* grown in batch culture with excess succinate and oxygen were performed by means of the intact cell assay, and 3 mol P_i were esterified per gram atom of oxygen respired.[22]

We do not accept the hypothesis that the identities of terminal oxidases determine the efficiency of oxidative phosphorylation and thereby influence the magnitude of the aerobic molar growth yield in *E. coli*. In this connection it is of interest that Jones has recently suggested that organisms containing cytochrome *d* possess an apparent selective advantage over those with cytochromes *o* or aa_3 during competitive growth at very low concentrations of dissolved oxygen.[27]

3. In still another instance of putative yield shortfall Poole and Haddock reported that continuous culture of *E. coli* at a dilution rate of 0.16 hr^{-2} during sulfate limitation in the presence of excess glycerol resulted in diminution of the aerobic molar growth yield as compared to values obtained when the culture was glycerol limited. Moreover, suspensions of cells harvested from sulfate-limited cultures displayed a smaller H^+/O value than did suspensions of cells obtained from unlimited or glycerol-limited cultures. The authors attributed these findings to loss of oxidative phosphorylation activity at "Site I" (NADH oxidation) consequent upon lowering of the amount of a sulfur-containing protein elicited by growth under sulfate-limited conditions.[50]

The essential control experiments necessary to establish an effect of sulfate limitation upon molar growth yield were not performed: variation of growth rate to establish the effect of that parameter upon yield, and estimation of the value of Y_{ATP}(max) as affected by sulfate limitation. Nevertheless, in agreement with Poole and Haddock, we have found that the value of Y_o (max) is diminished by sulfate limitation of continuous cultures in the presence of excess mannose, and that this is accompanied by an increment in the rate of maintenance respiration (Table 1). However, estimation of the value of Y_{ATP}(max) during anaerobic sulfate-limited continuous culture reveals that that parameter is also lowered by a significant amount (Table 1). The application of Equation 8 to the data ($r = 0.014$ mol ATP per gram and $\varrho = 0.0066$ mol of ATP per gram (see Table 2) shows that the value of Y_{ATP}(P) during sulfate-limited growth is only 11.7 g/mol ATP as compared with the value of 15.1 g/mol ATP observed with cultures limited by PTS sugars. Through Equations 9 and 10, as done previously with other substrates in Table 3, we find that the assumption that oxidative phosphorylation operates at a P/O value of 3 yields a P/O value calculated from our sulfate-limited cultures of 2.4 mol P_i esterified per gram atom of oxygen consumed, while the assumption that a P/O of 2 obtains leads to a value calculated from our results of 2.7 mol P_i esterified per gram atom oxygen consumed. As these values differ only insignificantly from those obtained from mannose-limited cultures expressed in Table 3, we conclude that no difference exists between the efficiency of oxidative phosphorylation during sulfate-limited growth in the presence of excess mannose and during mannose-limited growth in the presence of excess sulfate. Moreover, the values of Y_O(max) of continuous cultures supplied with succinate and limited by succinate, oxygen, phosphate, or sulfate vary only by small amount (Table 1) when the effects of maintenance respiration changes are taken into account. The decrement of aerobic molar growth yield consequent upon sulfate-limited growth as reported by Poole and Haddock can be most simply rationalized as due to an increase in the rate of maintenance respiration brought about by sulfate limitation, as we show in Table 1 for both succinate- and mannose-grown cultures.

For these reasons, we cannot accept the conclusion of Poole and Haddock that sulfate-limited growth affects the efficiency of oxidative phosphorylation as demonstrated by the change of areobic molar growth yield, even though the apparent efficiency of respiration-dependent proton extrusion is affected by such limitation.

We feel that no convincing demonstration of yield shortfall due to modification of the apparatus of oxidative phosphorylation has yet been made.

C. Aerobic Yield Shortfall Due to Oxygen Utilization by Nonrespiratory Means

During the aerobic metabolism of certain substrates by bacteria, oxygen consumption may occur due to reactions not related to terminal respiration; hence the oxidation reactions involved do not drive oxidative phosphorylation. Growth of pseudomonads at the expense of catechol as source of carbon and energy, for example, requires that the six-membered ring be opened by a dioxygenase-catalyzed reaction leading to the incorporation of oxygen atoms into the oxidized product.[6] Methane oxidation by methylotrophs involves the hydroxylation of the substrate through a reaction catalyzed by a mixed-function oxidase, with the incorporation of atmospheric oxygen into the product, methanol. The simple precaution of reporting aerobic molar growth yields only after correction for oxygen consumption occurring through such reactions is strongly recommended so as to avoid confusion in the interpretation of results. None of the aerobic yield estimations using *E. coli* as reported in Table 1 have been performed with substrates whose metabolism requires significant oxygen consumption other than by terminal respiration.

A less clearly defined consideration of the influence of oxygen metabolism upon aerobic growth yield has been raised by Hassan and Fridovich, who demonstrated that the activity of three isoenzymes with superoxide dismutase (SOD) activity in *E. coli* K-12 increased with growth rate during aerobic glucose-limited growth in continuous culture.[17] Growth at the expense of carbon sources other than glucose led to an increase of SOD activity and Paraquat (methyl viologen) relieved the suppressive effect of glucose upon SOD synthesis, although supplementation of the growth medium with cAMP did not so serve. The authors suggested that the intracellular level of the superoxide radical determines the rate of synthesis of SOD, and that glucose-grown cells contain smaller amounts of SOD because " . . . the fermentative catabolism of glucose leads to a lower rate of production of this radical *than does the predominantly* oxidative catabolism of nitrogenous compounds and organic acids."[18] In support of this allegation no assays of carbon-containing products were presented. Our calculation of aerobic molar growth yield from their estimations of the specific respiratory rate as a function of growth rate by means of Equation 5 provides a value of $Y_O(max)$ of about 40 gram per gram atom of oxygen and of m_O of about 6.5 mg atoms O per gram per hour. The level of SOD activity appears to have no effect upon the efficiency of respiratory-dependent biomass formation, and therefore cannot account for the earlier results of Hempfling in which an effect of growth in the presence of glucose upon the apparent efficiency of growth was noted (see previous section).

In summary, we aver that the evidence for aerobic yield shortfall brought about by modification of either the operation ("slip reactions") or composition ("repression") of the oxidative phosphorylation apparatus is not convincing. We suggest instead a model of respiratory energy conservation in growing *Escherichia coli* in which oxidative phosphorylation functions with full efficiency, (i.e., P/O value about 3) under all circumstances, but in which the effective availability of ATP for the formation of biomass may be diminished according to the peculiarities of growth rate-dependent net transport or metabolism of the major available nutrients, especially the source of carbon and energy. The yield shortfall manifested by cultures growing at the expense of succinate ought to be accounted for when we learn the true energetic costs of succinate transport, which have probably been underestimated in the calculations presented in Table 2 (value of *r* for succinate cultures). Moreover, although the effect of formate upon the yields of mannose-limited cultures is not understood, we suspect that interruption or modification of the ion current at the membrane brought about by establishment of an equilibrium distribution of formate and formic acid within and without the cell is at the root of formate-elicited yield shortfall. At least part of the yield shortfall of succinate cultures and the modest yield shortfall of mannose *plus* acetate cultures may be attributable to a similar phenomenon common to all mono- and dicarboxylic acids that permeate the bacterial cell.

VI. COMPARISON OF THE THEORETICAL AND PRACTICAL VALUES OF $Y_{ATP}(P)$

During anaerobic growth of *Escherichia coli* limited by PTS sugars and galactose, the value of Y_{ATP} (P) of the cultures is only about half of that calculated using information about energy required for synthesis from preformed monomers of protein, RNA, DNA, lipids, and polysaccharides in amounts occurring in the bacterial cell.[62] This discrepancy is serious enough so that an explanation for its existence must be sought. Stouthamer and co-workers, using continuous culture techniques, have found a few examples of organisms and anaerobic growth conditions in which $Y_{ATP}(max)$ values are expressed that approach the theoretical maximum of about 30 g/mol ATP.[10,63] Rittenberg and Hespell have reported that the value of Y_{ATP} of *Bdellovibrio bacteriovorus* during aerobic intraperiplasmic growth (a condition in which a full complement of preformed monomers might be expected to be available) also approaches 30 g/mol ATP.[56] The latter authors did not examine the effect of varying the steady-state growth rate upon molar growth yield, so that it is not possible to evaluate their results.

Several hypotheses might be offered to account for the discrepancy between theoretical and practical values of $Y_{ATP}(P)$.

1. The calculations may have underestimated the amount of energy necessary to drive certain polymerization reactions. Among other considerations, developments in our understanding of the intimate details of protein and nucleic acid synthesis must be closely followed with this notion in mind.

2. Additional, unanticipated energy-requiring reactions may exist, such as the growth rate-dependent expenditure of energy to distribute asymmetrically the various phospholipid classes within the cell membrane, transport of monomers between intracellular compartments for the synthesis of peptidoglycan or periplasmic constituents, or metabolically essential phosphorylation and dephosphorylation of membrane proteins.

3. "Energy-spilling" reactions may exist that are not influenced by the identity of the energy source or the growth-limiting substrate and which therefore are in operation under all growth conditions.

Stouthamer and Bettenhaussen determined some yield and maintenance properties of an ATPaseless mutant (M2-6) of *E. coli* K-12 in continuous culture.[64] The value of $Y_{ATP}(max)$ of the parent strain was 8.5 g/mol ATP during glucose-limited anaerobic continuous culture. Since M2-6 is incapable of normal anaerobic growth, Stouthamer and Bettenhaussen estimated the values of $Y_0(max)$ of aerobic glucose-limited and ammonium-limited continuous cultures (17.5 and 9.2 g/g atom oxygen, respectively) and then inferred values of $Y_{ATP}(max)$ by means of the stochastic model put forth by de Kwaadsteniet, et al.[9] and the assumption that the P/O value was zero. These calculations produced estimates of $Y_{ATP}(max)$ of 17.6 and 20.0 g/mol ATP for glucose- and ammonium-limited cultures, respectively.

The apparent increment of values of $Y_{ATP}(max)$ of the ATPaseless mutant as compared to that of the parent strain was rationalized by the investigators as possibly being due to a leak of protons linked in some manner to ATP hydrolysis in the parent. In the absence of the membrane ATPase, proton leakage would not occur in conjunction with ATP hydrolysis in the mutant. It was estimated that about half of the ATP generated during the parent's catabolism was utilized for "membrane energetization" consequent upon proton leakage, but that respiration without ATP formation was capable of supplying energy for such membrane processes in the mutant. That is, relatively more ATP was available for biosynthetic purposes in the mutant than in the parent. In support of this hypothesis the authors calculated that the P/O value of the aerobic cultures of the parent was 1.46 mol P_i esterified per g atom oxygen consumed. Because yields were measured at only two growth rates, the calculation of the P/O value was made with the unwarranted assumption that the rates of energy expenditure for maintenance were identical in the anaerobic and aerobic glucose-limited continuous cultures of the parent. Stouthamer and Bettenhaussen conclude " . . . that a relatively large amount of energy is required for membrane energetization and that this is the main task of respiration. . . . ".

We find the interpretation of their results by Stouthamer and Bettenhaussen to be unsatisfactory. First, we question the assumption that no ATP is formed through oxidative phosphorylation in cultures of the mutant M2-6. The method of assay of oxidative phosphorylation in intact cells employed by them, which measures the changes of levels of adenine nucleotides for only a few seconds after adding oxygen to anaerobic cells, is not capable of detecting synthesis of small amounts of ATP dependent upon respiration. Even a small amount of oxidative phosphorylation during growth on glucose, although insufficient to permit growth at the expense of a substrate metabolized without substrate level phosphorylation, would introduce a serious error in the calculation of the value of $Y_{ATP}(max)$ by the expression of de Kwaadsteniet, et al. Second, our calculation of the value of $Y_0(max)$ of the parent strain from the few data supplied by Stouthamer and Bettenhaussen reveals that 31.2 g/g atom oxygen is formed at a rate of maintenance respiration of about 0.7 mg atoms 0 per gram per hour, values commensurate with those found by us in mannose-limited *E. coli* B (Table 1). After correction for substrate level phosphorylation the apparent P/O value of the parent strain is about 3.3 mol P_i esterified per g atom of oxygen consumed.

In spite of these criticisms the apparent increase of Y_{ATP}(max) values consequent upon complete or partial loss of membrane ATPase activity must be seriously considered. At least part of the increment of yield ought to be due to increases of coupling efficiency between respiration and energy-dependent processes that are driven by membrane electrical potential or osmotic gradients arising due to elimination of intermediate ATP formation. As the capacitance of the membrane is small, cells capable of oxidative phosphorylation would be expected to synthesize ATP whenever possible; ATP would then be a major source of energy to drive membrane potential-dependent reactions, even though respiration is capable of supplying that energy directly. The number of individual, but sequential reactions linking respiration to generation of the membrane potential *via* ATP synthesis and hydrolysis is not known, but a sequence of only four reactions (for example, and in temporal sequence, proton extrusion, proton relaxation coupled to ATP formation, ATP hydrolysis coupled to generation of membrane potential, and application of membrane potential to a given membrane-specific task), each 90% efficient, will result in a loss of one third of originally conservable energy liberated through respiration and applied to membrane functions, while the same sequence of reactions, each 85% efficient, will dissipate nearly half of that energy. By eliminating the phosphorylation/dephosphorylation route of respiratory energy conservation, direct and hence more efficient coupling between respiration and supply of energy to maintain the required membrane potential is permitted ATPaseless mutants.

We feel that the discord between theoretical and practical values of Y_{ATP}(P) is less a consequence of unknown energy-consuming reactions than of the kinetic structure of coupling of conserved energy to its expenditure in biosynthetic reactions. All previous calculations of the value of Y_{ATP}(P) have ignored the phenomenon of imperfect coupling efficiency through linked reactions by assuming strict integral stoichiometries of energy consumption, a notion questionable on thermodynamic grounds. Because the actual overall coupling efficiencies are not known, however, we have persisted in using integral values for the stoichiometry of energy expenditure in our calculations of the values of r and ϱ as expressed in Table 2. The "theoretical" values of Y_{ATP}(P) are hence the consequences of inadequate theories, and in that sense, it is surprising that any examples of agreement between "theoretical" and practical molar growth yields have been realized. Of the three categories of hypotheses to explain Y_{ATP}(P) "shortfall" offered earlier in our discussion, the first is therefore the most significant, although the second and third possibilities cannot be eliminated with the scanty information at hand.

VII. MAINTENANCE METABOLISM

The growth rate-independent term of Equation 7, or rate of substrate utilization for maintenance purposes, is an important parameter influencing the practical molar growth yield of a microbial culture, especially when organisms are grown at low growth rates. Inspection of Table 1 shows that *Escherichia coli* manifests a wide range of rates of growth rate-independent energy utilization varying with the identity of the carbon source, the identity of the growth-limiting nutrient, and the presence or absence of oxygen. During anaerobic growth the rate of ATP utilization for maintenance purposes varies from about 6 to about 20 mmol ATP per gram per hour. The variation of rates of maintenance respiration from about 1.7 to about 25 mg atoms 0 per gram per hour is even greater than the range of anaerobic maintenance expenditures and the calculation of actual energy expenditure through the use of estimated P/O values (Table 3) extends the range of differences from about 4 mmol ATP per gram per hour (gluconate) to about 74 mmol ATP per gram per hour (glutamate).

The pattern of variation of rates of maintenance metabolism as the identity of the carbon source is changed differs from that of the variation of values of $Y_{ATP}(max)$ and $Y_0(max)$. Where the identity of the carbon source predominantly determines the magnitude of the molar growth yield, independent of the nature of the growth-limiting nutrient, the opposite appears to be the case with regard to the magnitude of the rate of maintenance metabolism.

In an attempt to rationalize these variations we will first consider those reactions that are likely to be the sites of maintenance energy, and then develop a model accounting for the variation of the magnitude of maintenance metabolism with the identity of the growth-limiting nutrient.

A. Identifying Sites of Energy Expenditure for Maintenance Purposes

We restrict the definition of "maintenance" to those energy-requiring reactions essential for growth and survival that do not directly result in biomass formation. With Pirt (Equation 4), we moreover require that the rate of energy expenditure in a maintenance process be independent of the rate of growth. According to this definition, likely reactions determining the magnitude of maintenance metabolism are turnover of macromolecules, motility (when applicable), osmotic regulation, homeostasis of the intracellular chemical composition, and other processes acting in opposition to "The thermodynamic and kinetic instabilities of many biochemical molecules . . . ."[30]

1. Turnover of Macromolecules

Pine has described rates of protein turnover of 2.5%/hr and 3.0%/hr in *E. coli* B during batch growth in glucose-containing ($\mu = 0.84$ hr^{-1}) and in acetate-containing medium ($\mu = 0.25$ hr^{-1}), respectively.[48] With different methods, Nath and Koch[43] identified a pool of protein comprising 4 to 8% of total cellular protein with a half-life of about 60 min. These two independent determinations are in reasonable agreement, and will be taken as the basis of our estimate. Assuming a requirement of 45 mmol of ATP to synthesize 1 g of protein,[11] protein turnover requires between 0.55 and 1.1 mmol ATP per gram per hour. This is likely to be an underestimate, as an energy requirement for protein degradation has been reported to occur both in vivo and in vitro (see Reference 12, pp. 787-791).

Ribonucleic acids are the second most abundant polymer in bacteria, and estimates of the turnover rate of ribosomal RNA, the major class of RNA molecules (75 to 80% of cellular RNA), suggest that between 1 to 5% of that fraction turns over per hour.[36] Assuming that transfer RNA turns over at a similar rate, RNA turnover would require from 0.25 to 0.13 mmol ATP per gram per hour. (The turnover of messenger RNA, the most labile species, is already included in the estimate of energy required for protein turnover.)

The contribution of peptidoglycan turnover to the overall maintenance requirement appears to be dependent upon the organism considered. In *Lactobacillus acidophilus* about one third of the complement of peptidoglycan turns over per generation, while in *E. coli*, *Salmonella typhimurium* and *Streptococcus faecalis* very little turnover has been detected.[2,65]

Significant turnover of fatty acids in growing *E. coli* has not been found, and the glycerol and phosphate moieties of phosphatidylethanolamine, the predominant cellular phospholipid, have also been shown to be stable (see Reference 5, pp.47—48).

We conclude that the cost of total macromolecular turnover in growing *E. coli* is between 0.6 and 1.2 mmol. ATP per gram per hour, and that the rates of the reactions comprising that maintenance requirement show little, if any, dependency upon the growth rate of growing bacteria.

2. Motility

An estimate of the amount of ATP required for motility, based on the work needed to overcome the viscous drag of water on a typical bacterium (0.5×2.0 μm) swimming at a rate of 50 μm/sec, is about 4.4×10^{-10} erg/cell/sec or 1.5×10^{-4} kcal/g/hr.[41] Assuming an efficiency of 4 kcal of mechanical work per mole ATP hydrolyzed,[3] the maintenance requirement to support motility is about 0.4 mmol ATP per hour. This estimate does not include energy required to replace flagella that are detached and lost into the surrounding culture fluid. In any event, the isolate of *E. coli* B used in our investigations is not flagellated; hence this maintenance requirement does not enter into the present calculations.

3. Chemical and Osmotic Regulation

The most abundant cation in most bacteria is K^+, and reports of substantial apparent concentration gradients are found in the literature.[38] The maintenance of such gradients would seem to place a sizable demand upon the energy supply of the cell. Damadian and co-workers have obtained evidence, however, that the bulk of intracellular potassium ion is present as counterion neutralizing cellular macromolecules, which carry a predominantly negative charge (for review see Reference 7). According to this hypothesis the true activity gradient of potassium ion is considerably smaller than has been generally thought. Damadian's group has furthermore provided data that demonstrate that free intracellular K^+ ion serves an osmotic regulatory function in consort with an anion such as Cl^- ion, entering the cell in response to and driven by an osmotic potential. In agreement with these findings we conclude that the apparent gradients of K^+, Na^+ and Mg^{2+} ions observed during growth not limited by those cations do not require significant expenditure of energy for their maintenance.

Harold and Papineau have shown that proton movements across a biological membrane are electrogenic[16] and, accordingly, the proton is the most likely ion to be maintained in a true state of transmembrane disequilibrium. Proton movement across the membrane associated with transport of other solutes or generation of ATP, as predicted by the chemiosmotic hypothesis, ought to be without net effect on the maintenance term. Transmembrane movements of protons that are not associated with productive work lead to a loss of free energy, however, and may result in expenditure of energy appearing in the maintenance requirement. Based upon the measurements of Mitchell and Moyle using isolated mitochondria,[40] we assume nonspecific leakage to occur at a rate of 0.11 μg-ion H^+ per gram protein per unit pH-sec. If the electromotive force of 120 mv measured in *E. coli*[47] is assumed equivalent to a pH gradient of two units, then 360 μ equiv. H^+ per gram per hour is lost through imperfect proton impermeability of the membrane. Further assuming a stoichiometry of 2 equivalents H^+ per mole ATP, we arrive at an estimate of 0.2 mmol ATP per gram per hour as the maintenance requirement necessary to replace energy loss through proton leakage.

4. Leakage of Cellular Components

Assuming that growing bacterial cells contain identical concentrations of monomers (amino acids, nucleotides, metabolites) regardless of the growth conditions, and that these monomers leak from the cell at a constant rate, the energetic cost of replacement by synthesis from the carbon source or reabsorption by the cell ought to be a component of the maintenance requirement. Roberts, et al. have reported that *E. coli*, growing with glucose as carbon source with a generation time of 1 hr leaks about 2 to 4% of available nascent monomers per generation.[57] If this population of monomers is drawn from the overall cellular contents, the maintenance requirement would be increased by about 0.13 to 0.26 mmol ATP per gram per hour during growth at the expense of hexose. The maintenance load due to leakage would increase upon changing the carbon source according to the value of ϱ peculiar to the carbon source.

Little information is available about leakage of polymers during normal growth. Loss of a small fraction of a lipopolysaccharide-phospholipid-protein complex from the cell envelope has been described,[59] but no systematic study of the effect of changing growth conditions upon such excretions has been performed. Under certain conditions, limitation of cells supplied with excess carbohydrate by restricting the nitrogen source can lead to massive synthesis of extracellular polysaccharide.[44] Interpretation of yield estimations performed with such cultures should be made with trepidation.

The results of our calculations seeking a minimal value for the rate of energy expenditure to serve maintenance processes are summarized in Table 4. We conclude that the minimum value of m_{ATP} in *E. coli* is 0.9 to 1.7 mmol ATP per gram per hour. As this quantity accounts at best for only about one third of the minimum observed value, it is clear that we have underestimated minimal maintenance demands, either by neglecting reactions of which we are unaware or by assigning values for energy expenditure to known reactions that are too small.

This elementary analysis predicts no dependency of the rate of maintenance metabolism upon the composition of the growth medium. As inspection of Table 1 discloses, such a compositional dependency exists and that observation, along with the probable underestimate of minimal maintenance requirements, prompts the development of a model accounting for the available results and facilitating further experimentation.

B. A Model of Maintenance Metabolism

At least two hypotheses may be put forth to rationalize the strong dependency of the rate of maintenance metabolism upon the identity of the growth-limiting nutrient:

1.   When a given solute limits growth, energy is required for its uptake (i.e. "active" transport), but when it is present in excess, transport occurs through passive, energy-nondependent, routes.
2.   The mechanisms of "active" transport persist no matter what the concentration of available nutrient might be, but the amount of gross transport is greater at low external concentrations than at high extracellular levels.

We immediately reject the first hypothesis since its acceptance would predict changes in the value of $Y_s(\text{max})$ as well as m_s dependent upon the nature of the growth-limiting nutrient, and such is not the case with mannose- or succinate-limited aerobic cultures described in Table 1.

A mechanistic model rationalizing the second hypothesis will be developed with the aid of two assumptions:

1.   Even under conditions of growth limitation by a solute subject to "active" transport, that transport is a near-equilibrium reaction.
2.   Three processes affect the net rate of energy-dependent transport — the rate of uptake, the rate of leakage of solute from the cells in a manner independent of the rate of growth, and the rate of net metabolic transformation of the solute occurring in a growth rate-dependent manner, as in Equation 7.

The rate of change of concentration of a solute Z in a cell growing in the steady state is

$$d[Z]_{i/dt} = dZ/dt - P_m \cdot A \cdot [Z]_{in} + P_m \cdot A \cdot [Z]_{out} - \mu \cdot r' = 0 \quad (13)$$

Table 4
MINIMAL MAINTENANCE DEMANDS OF
ESCHERICHIA COLI DURING
UNRESTRICTED GROWTH IN THE
PRESENCE OF OXYGEN

Reaction	Energy expended (mmol ATP/g/hr)
Protein turnover	0.55—1.1
Ribosomal and transfer RNA turnover	0.02—0.13
Leakage of protons	0.2
Leakage of monomers	0.13—0.26
Total expenditure of ATP to serve minimal maintenance demands	0.9 —1.7 mmol/g/hr

where dZ/dt = the rate of active uptake of solute Z,

P_m = the membrane permeability coefficient of solute Z as calculated from Fick's first law,

A = membrane surface per unit biomass,

r' = moles of the solute Z metabolized for growth per unit biomass (see Equation 7).

The rate of active uptake of solute Z is therefore

$$dZ/dt = P_m \cdot A \cdot d[Z] - \mu \cdot r' \qquad (14)$$

where $d[Z]$ = $[Z]_{in} - [Z]_{out}$.

If k moles of phosphate anhydride bonds are hydrolyzed per mole of solute Z translocated, then the rate of hydrolysis for total transport activity is

$$d(\sim P)/dt = P_m \cdot A \cdot K \cdot d[Z] + \mu \cdot r \qquad (15)$$

where *r* is as defined in Equation 7.

The rate of energy expenditure in growth rate-independent processes may be defined as

$$d(\sim P)/dt = P_m \cdot A \cdot K \cdot d[Z] = \alpha M_s \qquad (16)$$

where M_s = the rate of utilization of energy-supplying substrate to counteract diffusive leakage,

and α = the stoichiometry of ATP produced per mole of energy-supplying substrate.

If the solute Z is available in amounts that serve to limit growth, then μ is dependent upon the extracellular steady-state concentration of Z, and d [Z] will vary with μ. Since $[Z]_{in}$ is in the millimolar concentration range for most major cellular solutes, however, and since $[Z]_{out}$ is in the micromolar range during conditions of growth limitation by Z, the dependency of d[Z] and αM_s upon μ will be negligible.

All cellular components ought to contribute to total energy expenditure as follows

$$\alpha M_s = \sum_{i \to j} P_{m_i} \cdot K_i \cdot d[Z_i] \cdot A \qquad (17)$$

Hence, if the values of P_m, K and d [Z] are determined for all cellular components, the value of total energy expenditure ought to be calculable. Although this expression assumes that the value of P_m in the forward direction equals the value of P_m for outward diffusion, these parameters may in fact be different, and the expressions should be adjusted accordingly. Moreover, evidence exists that suggests that some leakage may be permease mediated.[14,69] This eventuality does not significantly affect our analysis, as the equations may be easily modified to include the operation of such a mechanism.

We believe that the maintenance term in Pirt's expression (Equation 4) ought now to be redefined with these concepts as

$$m = m'_s + M_S \tag{18}$$

where m = the parameter as expressed in Equation 4,

m'_s = the "basal maintenance rate", that is, a term including those maintenance reactions that occur independently of the nature of the growth medium, such as macromolecular turnover, proton leakage, etc.

and M_s = the energy expenditure necessary to counteract the leakage of the growth-limiting nutrient.

Through this definition $m = m'_s$, when all components of the growth medium are available in such concentrations as to permit unrestricted growth (i.e., M_s = zero).

Inspection of Table 1 shows that succinate-limited cultures manifest a rapid rate of maintenance expenditure of ATP (about 12 mmol ATP per gram per hour), but m_{ATP} during oxygen limitation falls to less than 5 mmol ATP per gram per hour. Intermediate rates of maintenance expenditure of ATP are observed when succinate-containing continuous cultures are limited with phosphate or sulfate (about 7 to 8 mmol ATP per gram per hour). We consider that the oxygen-limited condition approximates the case in which the value of M_S of Equation 16 approaches zero, and that the increased maintenance requirement imposed by limitation with succinate, phosphate, or sulfate is brought about as a consequence of the leakage of those limiting nutrients due to the large concentration gradients imposed. Similarly mannose-limited cultures manifest a low rate of maintenance respiration, an expected outcome, since mannose is transported through group translocation and all other nutrients are present in excess. Limitation of continuous cultures supplied with excess mannose by restricting the supply of sulfate or phosphate again results in substantial increments of the rates of maintenance respiration and hence of rates of expenditure of ATP for maintenance of the necessary intracellular concentrations of sulfate and phosphate.

Of interest in this connection are the data of Rosenberg et al.[58] *E. coli*, grown on lactate and deprived of P_i can take up phosphate at a maximal rate of 5.2 mmol P_i per gram per hour. The intracellular P_i pool rapidly turns over, 70% exchanging with extracellular P_i in 1 min. A bacterium growing at $\mu = 1$ hr^{-1} requires only 0.7 mmol P_i per gram per hour to supply net biosynthetic demands. If these values are substituted into our expression, assuming a stoichimetry of 1 ATP per P_i transported, we arrive at a value of M_{ATP} of 4.5 mmol ATP per gram per hour. This value, added to the basal maintenance requirement, agrees well with the observed maintenance rate of phosphate-limited succinate cultures. The value of M_O of phosphate-limited mannose cultures is larger than would be predicted.

We summarize our considerations of yield and maintenance by substituting Equation 16 into Equation 7

$$1/Y_S = m'_s/\mu + M_S/\mu + r + \rho + 1/Y_S(P) \tag{19}$$

and urge that this expression and the concepts involved in its development be employed toward rationalizing the dependency of the magnitudes of microbial molar growth yield and maintenance metabolism upon the composition of the growth medium.

REFERENCES

1. **Bauchop, T. and Elsden, S.**, The growth of micro-organisms in relation to their energy supply, *J. Gen. Microbiol.*, 23, 457, 1960.
2. **Boothby, D., Daneo-Moore, L., Higgins, M. L., Coyette, J., and Shockman, G. D.**, Turnover of bacterial cell wall peptidoglycans, *J. Biol. Chem.*, 248, 2161, 1973.
3. **Brokaw, C. J.**, Adenosine triphosphate usage by flagella, *Science*, 156, 76, 1967.
4. **Burnell, J. N., John, P., and Whatley, F. R.**, The reversibility of active sulphate transport in membrane vesicles of *Paracoccus denitrificans*, *Biochem. J.*, 150, 527, 1975.
5. **Cronan, J. E., Jr. and Vagelos, P. R.**, Metabolism and function of the membrane phospholipids of *Escherichia coli*, *Biochim. Biophys. Acta*, 265, 25, 1972.
6. **Dagley, S., Evans, W. C., and Ribbons, D. W.**, New pathways in the oxidative metabolism of aromatic compounds by microorganisms, *Nature (London)*, 188, 560, 1960.
7. **Damadian, R.**, Cation transport in bacteria, *Crit. Rev. Microbiol.*, 2, 377, 1973.
8. **Dawes, E. A.**, 1976. Endogenous metabolism and the survival of starved procaryotes, in *Survival of Vegetative Microbes*, Gray, T. R. G. and Postgate, J. R., Eds., Cambridge Univ. Press, Cambridge 1976, 19.
9. **de Kwaadsteniet, J. W., Jager, J. C., and Stouthamer, A. H.**, A quantitative description of heterotrophic growth in micro-organisms, *J. Theor. Biol.*, 57, 103, 1976.
10. **de Vries, W., Kapteijn, W. M. C., van der Beek, E. G., and Stouthamer, A. H.**, Molar growth yields and fermentation balances of *Lactobacillus casei* L3 in batch cultures and in continuous cultures, *J. Gen. Microbiol.*, 63, 333, 1970.
11. **Forrest, W. W. and Walker, D. J.**, 1971. The generation and utilization of energy during growth, Rose, A. H. and Wilkinson, J. F., Eds., *Adv. Microb. Physiol.*, 5, 213, 1971.
12. **Goldberg, A. L. and St. John, A. C.**, 1976. Intracellular protein degradation in mammalian and bacterial cells: Part 2. *Annu. Rev. Biochem.*, 45, 747, 1976.
13. **Gunsalus, I. C. and Shuster, J.**, Energy-yielding metabolism in bacteria, in *The Bacteria*, Vol. 2, Gunsalus, I. C. and Stanier, R. Y., Eds., Academic Press, New York, 1961, 1.
14. **Halpern, Y. S., Barash, H., and Druck, K.**, Glutamate transport in *Escherichia coli* K-12: nonidentity of carriers mediating entry and exit, *J. Bacteriol.*, 113, 51, 1973.
15. **Harold, F.**, Membranes and energy transdution in bacteria, in *Curr. Top. Bioenerg.*, Sanadi, R., Ed., 6, 83, 1977.
16. **Harold, F. and Papineau, D.**, Cation transport and electrogenesis by *Streptococcus faecalis*. II. Proton and sodium extrusion, *J. Membr. Biol.*, 8, 45, 1972.
17. **Hassan, H. M. and Fridovich, I.**, Physiological function of superoxide dismutase in glucose-limited chemostat cultures of *Escherichia coli*, *J. Bacteriol.*, 13, 805, 1977.
18. **Hassan, H. M. and Fridovich, I.**, Regulation of superoxide dismutase synthesis in *Escherichia coli*: glucose effect, *J. Bacteriol.*, 132, 505, 1977.
19. **Hempfling, W. P.**, Repression of oxidative phosphorylation in *Escherichia coli* B by growth in glucose and other carbohydrates, *Biochem. Biophys. Res. Commun.*, 41, 9, 1970.
20. **Hempfling, W. P.**, Ed., *Microbial Respiration*, Dowden, Hutchinson, and Ross, Stroudsburg, Pa., 1970.
21. **Hempfling, W. P., and Beeman, D. K.**, Release of glucose repression of oxidative phosphorylation in *Escherichia coli* by cyclic adenosine 3′,5′-monophosphate, *Biochem. Biophys. Res. Commun.*, 45, 924, 1971.
22. **Hempfling, W. P. and Hertzberg, E. L.**, Techniques for measurement of oxidative phosphorylation in intact bacteria and in membrane preparations of *Escherichia coli*, in *Methods Enzymol.*, Packer, L. and Fleischer, S., Eds., 55 F, 164, 1979.
23. **Hempfling, W. P., Höfer, M., Harris, E. J., and Pressman, B. C.**, Correlation between changes in metabolite concentrations and rate of ion transport following glucose addition to *Escherichia coli* B, *Biochim. Biophys. Acta*, 141, 391, 1967.

24. Hempfling, W. P. and Mainzer, S. E., Effects of varying the carbon source limiting growth on yield and maintenance characteristics of *Escherichia coli* in continuous culture, *J. Bacteriol.*, 123, 1076, 1975.

25. Herbert, D., Ellsworth, R., and Telling, R. C., The continuous culture of bacteria: a theoretical and experimental study, *J. Gen. Microbiol.*, 14, 601, 1956.

26. Hertzberg, E. L. and Hinkle, P. C., Oxidative phosphorylation and proton translocation in membrane vesicles prepared from *Escherichia coli, Biochem. Biophys. Res. Commun.*, 58, 178, 1974.

27. Jones, C. W., Microbial oxidative phosphorylation, *Biochem. Soc. Trans.*, 6, 361, 1978.

28. Kashket, E. R. and Brodie, A.F., Oxidative phosphorylation in fractionated bacterial extracts. VIII. Role of particulate and soluble fractions from *Escherichia coli, Biochim. Biophys. Acta*, 78, 52, 1963.

29. Kjeldgaard, N. O. The kinetics of ribonucleic acid- and protein formation in *Salmonella typhimurium* during the transition between different states of balanced growth, *Biochim. Biophys. Acta*, 49, 64, 1961.

30. Mallette, M. F., Validity of the concept of energy of maintenance, *Ann. N.Y. Acad. Sci.*, 102, 521, 1963.

31. Mandelstam, J. and Halvorson, H., Turnover of protein and nucleic acid in soluble and ribosome fractions of non-growing *Escherichia coli, Biochim. Biophys. Acta*, 40, 43, 1960.

32. Mandelstam, J. and McQuillen, K., Eds., *The Biochemistry of Bacterial Growth*, 2nd ed., Blackwell Scientific Pub., Oxford, 1973.

33. Marr, A. G., Nilson, E. H., and Clark, D. J., The maintenance requirement of *Escherichia coli, Ann. N.Y. Acad. Sci.*, 102, 536, 1963.

34. Martin, G. A. and Hempfling, W. P., A method for the regulation of microbial population density during continuous culture at high growth rates, *Arch. Microbiol.*, 107, 41, 1976.

35. McAfee, R. D. and Nieset, R. T., Studies of amino acid metabolism in *Escherichia coli* with ^{15}N, *Biochim. Biophys. Acta*, 31, 365, 1959.

36. Meselson, M., Nomura, M., Brenner, S., Davern, C., and Schlessinger, D., Conservation of ribosomes during bacterial growth, *J. Mol. Biol.*, 9, 696, 1964.

37. Meyer, D. J. and Jones, C. W., Oxidative phosphorylation in bacteria which contain different cytochrome oxidases, *Eur. J. Biochem.*, 36, 144, 1973.

38. Minkoff, L. and Damadian R., Caloric catastrophe, *Biophys. J.*, 13, 167, 1973.

39. Mitchell, P., Coupling of phosphorylation to electron and hydrogen transfer by a chemi-osmotic type of mechanism, *Nature (London)*, 191, 144, 1961.

40. Mitchell, P. and Moyle, J., Acid-base titration across the membrane system of rat-liver mitochondria. Catalysis by uncouplers, *Biochem. J.*, 104, 588, 1967.

41. Morowitz, H., The energy requirements for bacterial motility, *Phys. Rev.*, 87 (Abstr. C4), 186, 1952.

42. Moss, F., The influence of oxygen tension on respiration and cytochrome a_2 formation of *Escherichia coli, Aust. J. Exp. Biol. Med. Sci.*, 30, 531, 1952.

43. Nath, K. and Koch, A. L., Protein degradation in *Escherichia coli*. II. Strain differences in the degradation of protein and nucleic acid resulting from starvation, *J. Biol. Chem.*, 246, 6956, 1971.

44. Neijssel, O. M. and Tempest, D. W., The regulation of carbohydrate metabolism in *Klebsiella aerogenes* NCTC 418 organisms, growing in chemostat culture, *Arch. Microbiol.*, 106, 251, 1975.

45. Neijssel, O. M. and Tempest, D. W., Bioenergetic aspects of aerobic growth of *Klebsiella aerogenes* NCTC 418 in carbon-limited and carbon-sufficient chemostat culture, *Arch. Microbiol.*, 107, 215, 1976.

46. Neijssel, O. M. and Tempest, D. W., The role of energy-spilling reactions in the growth of *Klebsiella aerogenes* NCTC 418 in aerobic chemostat culture, *Arch. Microbiol.*, 110, 305, 1976.

47. Padan, E., Zilberstein, D., and Rottenberg, H., The proton electrochemical gradient in *Escherichia coli* cells, *Eur. J. Biochem.*, 63, 533, 1976.

48. Pine, M., Steady-state measurement of the turnover of amino acids in the cellular proteins of growing *Escherichia coli:* existence of two kinetically distinct reactions, *J. Bacteriol.*, 103, 207, 1970.

49. Pirt, S. J., The maintenance energy of bacteria in growing cultures, *Proc. R. Soc. London, Ser. B.*, 163, 224, 1965.

50. Poole, R. K. and Haddock, B. A., Effects of sulphate-limited growth in continuous culture on the electron-transport chain and energy conservation in *Escherichia coli* K12, *Biochem. J.*, 152, 537, 1975.

51. Postma, P. W. and Roseman, S., The bacterial phospho-*enol* pyruvate:sugar phosphotransferase system, *Biochim. Biophys. Acta*, 457, 213, 1976.

52. Ramos, S., Schuldiner, S., and Kaback, H. R., The electrochemical gradient of protons and its relationship to active transport in *Escherichia coli* membrane vesicles, *Proc. Natl. Acad. Sci. U.S.A.*, 73, 1892, 1976.

53. **Rhoads, D. B. and Epstein, W.**, Energy coupling to net K⁺ transport in *Escherichia coli* K-12, *J. Biol. Chem.*, 252, 1394, 1977.

54. **Rice, C. W. and Hempfling, W. P.**, Oxygen-limited continuous culture and respiratory energy conservation in *Escherichia coli, J. Bacteriol.*, 134, 115, 1978.

55. **Rice, C. W., Cooper, N. L., and Hempfling, W. P.**, Unpublished experimental results.

56. **Rittenberg, S. C. and Hespell, R. B.**, Energy efficiency of intraperiplasmic growth of *Bdellovibrio bacteriovorus, J. Bacteriol.*, 121, 1158, 1975.

57. **Roberts, R. B., Abelson, P. H., Cowie, D. B., Bolton, B. T., and Britten, R. J.**, *Studies of Biosynthesis in Escherichia coli,* Carnegie Institute of Washington, Washington, D.C., 1963, 184.

58. **Rosenberg, H., Gerdes, R. G., and Chegwidden, K.**, Two systems for the uptake of phosphate in *Escherichia coli, J. Bacteriol.*, 131, 505, 1977.

59. **Rothfield, L. and Pealman-Kothenko, M.**, Synthesis and assembly of bacterial membrane components. A lipopolysaccharide-phospholipid-protein complex excreted by living bacteria, *J. Mol. Biol.*, 44, 477, 1969.

60. **Senez, J.**, Some considerations on the energetics of bacterial growth, *Bacteriol. Rev.*, 26, 95, 1962.

61. **Stevenson, R. and Silver, S.** Methylammonium uptake by *Escherichia coli:* evidence for a bacterial NH⁺₄ transport system, *Biochem. Biophys. Res. Commun.*, 75, 1133, 1977.

62. **Stouthamer, A. H.** A theoretical study on the amount of ATP required for synthesis of microbial cell material, *Antonie van Leeuwenhoek; J. Microbiol. Serol.*, 39, 545, 1973.

63. **Stouthamer, A.H. and Bettenhaussen, C. W.**, Energetic aspects of the anaerobic growth of *Aerobacter aerogenes* in complex medium, *Arch. Microbiol.*, 111, 21, 1976.

64. **Stouthamer, A.H. and Bettenhaussen, C. W.**, A continuous culture study of an ATPase-negative mutant of *Escherichia coli, Arch. Microbiol.*, 113, 185, 1977.

65. **Van Tubergen, R. P. and Setlow, R. B.**, Quantitative radioautographic studies on exponentially growing cultures of *Escherichia coli.* The distribution of parental DNA, RNA, protein, and cell wall among progeny cells, *Biophys. J.*, 1, 589, 1961.

66. **Vorisek, J. and Kepes, A.**, Galactose transport in *Escherichia coli* and the galactose-binding protein, *Eur. J. Biochem.*, 28, 364, 1972.

67. **Wagner, C., Odom, R., and Briggs, W. T.**, The uptake of acetate by *Escherichia coli* W, *Biochem. Biophys. Res. Commun.*, 47, 1036, 1972.

68. **Watson, T. G.**, Effects of sodium chloride on steady-state growth and metabolism of *Saccharomyces cerevisiae, J. Gen. Microbiol.*, 64, 91, 1970.

69. **Winkler, H. H. and Wilson**, The role of energy coupling in the transport of beta-galactosides by *Escherichia coli, J. Biol. Chem.*, 241, 2200, 1966.

Chapter 5

GENETIC STUDIES USING CONTINUOUS CULTURE*

Peter H. Calcott

One of the advantages of using continuous culture is that the cell's environment can be rigidly controlled. With this controlled environment, the cells maintain a constant reproducible physiology that is not continually changing, as in batch culture. Thus, unlike a batch culture, a continuous culture becomes essentially a time independent apparatus. Sometimes, this has been shown to be untrue. For instance, Tempest[1] reported a hysteresis effect on RNA content when cells were grown at progressively higher growth rates than progressively lower ones. Not only RNA content, but other macromolecules and metabolic activities showed a hysteresis effect.[1] Clearly, then continuous cultures are not always time independent.

This is not the only observation of variation in the physiology of cells grown in continuous culture. Before illustrating examples of these changes, it is necessary to examine some theory of why changes might or should occur.

Early microbiology relied on a technique for isolation of organisms from a sample, called enrichment culture. This involved inoculating the sample into a culture medium containing nutrients, and providing physical and chemical conditions that would give the organisms of interest a selective advantage. It allowed the "rapid" multiplication of the desired organism, but "repressed" the multiplication of the nondesired. The same types of selective procedures have been successfully used by microbial geneticists in isolating mutants. Since (in these selection cultures) the concentration of nutrients is rarely limiting, at least when the culture is started, selection of the desired cell type is based on the maximum growth rate (μ_{max}) of the cell.

Selection in continuous culture, not only between different organisms (see other chapters in this volume), but also between parents, variants, and mutants, is dependent not on μ_{max}, but on another parameter. It is true that μ_{max} would ultimately govern which cell type survived, but this μ is governed by the affinity of the cell for the substrate (K_s). This term is analoguous to the constant K_m of an enzyme obeying Michaelis-Menten kinetics. The growth rate of the cell is governed by the equation:

$$\mu = \mu_{max} \left(\frac{S}{K_s + S} \right)$$ where μ and μ_{max} are growth rates when substrate concentration is S and saturating, respectively;

K_s is substrate concentration when $\mu = \frac{1}{2}\mu_{max}$

Thus if two cells are together in continuous culture, the one with the lowest K_m or highest affinity for the growth-limiting substrate would be selected for. This can be seen more easily in Figure 1. At dilution rates (D) less than X, organism 1 will be selected for, while at D values greater than X, 2 will be selected. At X, both cell types will theoretically be maintained. Thus the two cell types may be of different species or be of the same strain, i.e., variant and/or mutant. Needless to say, the mutant may be introduced intentionally or be produced *in situ;* thus, a continuous culture is selective for an organism that is "more adapted" to the environment. In addition, this selection is enhanced since there is a flow of biomass from the culture vessel.[1] Tempest thus recommends that chemostats should not be run for periods longer than 4 weeks,

* This chapter was submitted June 1978.

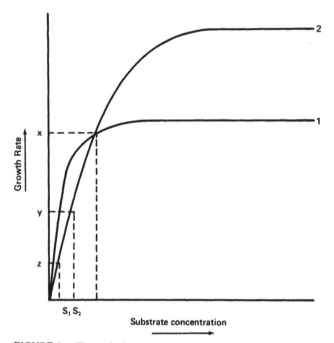

FIGURE 1.   Theoretical saturation curves for 2 organisms grown in
separate but identical chemostats. At a particular dilution rate (Y),
the concentration of growth limiting substrate in the chemostat would
be S_1 for organisms 1 and S_2 for organisms 2. Thus if both cell types
were mixed and the dilution rate set at Y both organisms would utilize
the substrate until S_1 was attained. At which point, growth rates of
the two cell types (1 and 2) would be Y and Z respectively. Thus organism 1 which is growing at a higher growth rate (μ) would be selected for and establish itself. Organism 2 would be at a selectively
disadvantage and be washed out of the culture, being unable to maintain μ = Y. At growth rates greater than X, organism 2 would out
compete 1.

and should never be used to inoculate another. This could be advocated if the operator
is not interested in selection. The use of continuous culture for selection has been
demonstrated on several occasions, notably by Novick and Szilard,[2,3] Strange and
Hunter,[4] Bryson,[5] and Munson and Bridges.[6]

The fate of the contaminant organisms is not always governed by the K_m alone.
Other factors that influence the fate of the organisms are the culture pH, temperature,
nutritional requirements of the organism, and composition of the medium. In addition, if the organisms interact (other than simple competition), the predictions of the
outcome become difficult to calculate.

Selection in chemostat of a variant over a parent has been observed by Lee and
Calcott.[40] They showed that the parent, *Escherichia coli*, when inoculated into a carbon-limited chemostat culture and grown at a D = 0.2 h^{-1} and 37°C was displaced by
a variant produced *in situ* in 3 to 4 weeks. The parent and variant were distinguishable
by colonial morphology differences on a complex medium (nutrient agar, casamino
acids, and yeast extract). The variant was also more susceptible to freezing and thawing
in water or saline than the parent. When grown under nitrogen-limitation, the parent
outgrew the variant. The precise nature of the parent to variant transition was not
understood.

Another demonstration of parent-variant selection has been described in detail by

Strange and Hunter.[4] In this case, the parent and variant of *Klebsiella aerogenes* differed in biochemical, immunological, chemical composition, and survival properties, susceptibility to substrate-accelerated death,[7] and colonial morphology. When approximately equal numbers of parent and variant were mixed in a chemostat culture (glycerol-limited), the variant took over the culture at D values of $0.5h^{-1}$ or below. At D $= 0.62h^{-1}$ the parent took over (Figure 2). Once again the nature of the genetic difference between parent and variant was not alluded to.

These parent-variant transitions are not defined systems genetically. A better example of a genetic selection occurring in continuous culture is that of strains producing hyper levels of β-galactosidase. It has been observed in *E. coli* and *K. aerogenes*. In both systems, when cells were inoculated into a lactose-limited chemostat and a steady state attained, a characteristic specific activity was attained for the enzyme β-galactosidase. On prolonged culture, the specific activity increased to a new maximum[8-13] (Figure 3). In some of the early reports, the initially inducible strain of *E. coli* was replaced by a constitutive strain.[8,11,12] Evidently at the low levels of lactose in the chemostat, constitutive cells produced higher levels of the enzyme β-galactosidase than the inducible strain. With a higher capturing rate of limiting substrate, the former was selected for. If, however, the competition was conducted in batch culture where high lactose concentration prevailed, the inducible strain outgrew the constitutive strain. Once the constitutive had been selected, selection for strain containing higher levels still continued. These hyper-level producers would only be maintained in continuous culture. In batch culture they were selected against and the cultures reverted to the characteristic constitutive strains. In addition, the hyperproducers were also very sensitive cells. When these hyperproducers were plated on a lactose medium, the cells failed to grow.[14] It is interesting to note that constitutive strains were not detected by Smith and Dean[9] and Calcott and Postgate[10] when they grew hyperproducing cultures of *K. aerogenes*. These hyperproducing strains of Novick and co-workers[8,11,12] produced an enzyme with similar K_m, molecular weight, ultra centrifuge sedimentation pattern, thermal stability, and immunological characteristics to that of the inducible strain. Thus the hyperproducers were not producing a better enzyme, but more. The question asked is "how does the cell produce more enzymes?" Clearly, there are several possible explanations. One is that the genes for β-galactosidase are derepressed further in the hyperproducers than the normal inducible strain. Alternatively, multiple copies of the genes for β-galactosidase are produced. A further possibility was alluded to by Clarke.[15] He favors that a mutation in the promoter might allow an increased rate of transcription of *lac* DNA. This latter model seems unlikely, since one would predict, not only a high rate of β-galactosidase production, but also higher levels of *lac*-permease and transacetylase, which are coordinately synthesized, unless polarity effects prevailed. High levels of permease are not produced.[9,10] The fate of transacetylase activity was not reported.

The genetic nature of the hyperproducing strains has been characterized. Novick and Horiuchi[12] have mapped the hyperproducing region of the *E. coli* chromosome. It appears to be located in the same general region as the *lac*-operon.[12] They also produced evidence that these "hyperproducing genes" were not associated with an F-factor.[12] They favor that several copies of the gene for β-galactosidase are present. Transduction experiments indicated that these extra genes were located very close to the *lac*-region.[12]

Smith and Dean[9] and Calcott and Postgate[10] do not favor the genetic nature of hyperproduction advocated by Novick and co-workers.[11,12] They favor that the levels are more of a reflection on the levels of repression and derepression that the cultures are experiencing. When a chemostat culture is carbon-limited, the level of carbon

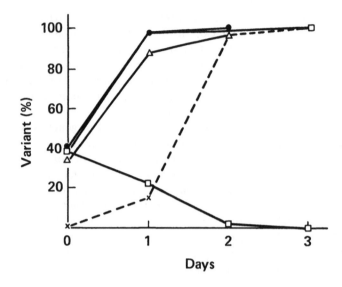

FIGURE 2.   Growth of glycerol-limited parent and variant strains of *Aerobacter aerogenes* in mixed culture in chemostats at various dilution rates. Samples of cultures of the organisms growing separately in chemostats at the given dilution rate were mixed in a third chemostat and the proportion of each organism was determined by differential colony counts after intervals of growth at 37°C. Proportion of varient organisms at dilution rates of 0.25 (a,X), 0.3 (△) 0.5 (•) and 0.62 (□) hr⁻¹. One mixed culture (x) was seeded with 0.1% variant. (From Strange, R. E. and Hunter, J. R., *J. Gen. Microbiol.*, 44, 255, 1966. With permission.)

source left unutilized in the medium is dependent on the growth rate of the culture and hence dilution rate. At low growth rates, the extracellular carbon concentration is vanishingly small, while at higher D values the concentration of the carbon source (whie still low) is higher. Paralleling this low intracellular concentration of carbon source is the intracellular concentration of carbon intermediates. Under lactose-limitation, the concentration of lactose and its breakdown products, galactose and glucose, will be very small intracellularly. Thus under induction situations, the final level of β-galactosidase will be a reflection of the level of induction of the *lac*-operon. This is dependent on an adequate supply of co-factors such as cAMP and cAMP binding protein. The presence of the former is dependent on the absence of glucose and its breakdown products. Thus selection for hyperproducing strains may be a selection for a strain producing higher levels of cAMP or other control nucleotides or selection for a strain with less susceptibility to catabolite repression. It would be interesting to know if the intracellular levels of cAMP are altered in their hyperproducers over normal strains. In addition, the properties of adenylate cyclase, particularly its glucose sensitivity, would be interesting.

Hegeman[16] has also been able to select for constitutive mutants in *Pseudomonans putida*. He used cultures limited by mandelate or benzylformate. Under these conditions, the early enzymes were constitutive, while the later enzymes were repressed in non-inducing situations. These mutants could be induced to higher levels than the parents by compounds that were ordinarily ineffective with the parent. Interestingly, starvation for the carbon *cum* energy source also induced the enzymes.

Continuous culture has also been used to select for dispensable enzymes. Cocito and Vogel[17] grew *E. coli* in a glucose-salt medium turbidostat. They added ornithine (a

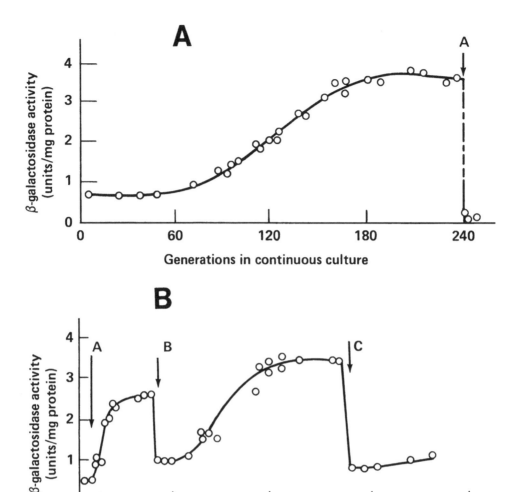

FIGURE 3. β-galactosidase activity of *Klebsiella aerogenes* in chemostat culture. A. An inoculum of *Klebsiella aerogenes* from a lactose-limited batch culture was grown continuously, D = 0.23 h⁻¹ at 37°C aerobically. β-galactosidase activity was determined on toluenized preparations once steady state had been attained. Initially, the culture was lactose-limited (2 g/l) but after 240 generations the medium was changed to a glycerol-limiting one (2 g/l) at A. B. A chemostat was set up as for A except that the inoculum was taken from the chemostat in A after 70 generations and the inflow medium contained 0.5 g lactose/l initially. The medium was latered at A to retain limiting lactose limitation but with a gratutitious inducer present, thio methyl galactoside (0.5 g lactose/l + 1.5 mM-TMG). At B, normal lactose-limiting conditions (0.5 g/l) were restored; at C, the medium was again altered to produce lactose-utilizing, ammonium-limiting conditions (10 g lactose/l + 6 mM-NH₄Cl). (From Calcott, P. H. and Postgate, J. R., *J. Gen. Microbiol.*, 85, 85, 1974. With permission.)

repressor of acetylornithinase synthesis) and found the enzyme levels in the culture were reduced 90% in some cultures, while others showed production of the enzyme whose synthesis was not repressed by ornithine.

Munson and Bridges[6] used a turbidostat to examine the reversion of trp⁻ *E. coli* to prototrophy induced by γ rays. Occasionally they found that after an initial rise in reversion, a second larger increase in frequency was noted (Figure 4). Further experiments indicated that this rapid second reversion was due only to a small proportion

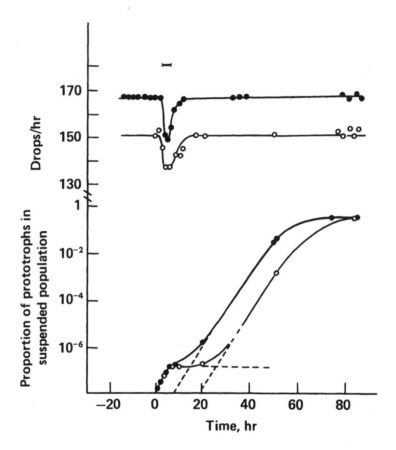

FIGURE 4.   Increases in prototroph frequency ("take-over") in two continuous cultures of *E. coli* WP2 trp⁻. The cultures were simultaneously exposed to
4500 rads of γ-radiation during the period 0 to 4 hr. The broken line indicates
the contribution from normal prototrophs and the dotted lines the contributions
from the hypothetical variant "sticky" prototrophs, their slopes being 0.36 hr⁻¹.
The growth rates (drops/hr) shown in the upper graphs show no changes except
during the irradiations (Munson and Bridges, 1964). (From Munson, R. J. and
Bridges, B. A., *J. Gen. Microbiol.*, 37, 411, 1964. With permission.)

of cells in the population. In addition, wall growth that was prominent contained these
"rapid reverting" cells at a much higher concentration than the culture as a whole.
These workers hypothesized that both trp⁻ and trp⁺ cells became attached to the culture
vessel wall. The chance of attachment and detachment was constant. Once a steady
state was attained, a "sticky" mutant prototrophe became attached. The attachment
rate was constant, but the rate of detachment was severely reduced in this prototrophe.
At this time there would be "take over" of the surface by this mutant.

Continuous cultures have been used extensively to study the incidence of antibiotic
resistance in bacteria. Munson[18] has reviewed several experiments by various workers.

Bryson[5] used a turbidostat with an extra nutrient supply, called a "turbidostatic
selector". He was able to select for *E. coli* mutants capable of growth at concentrations
of neomycin 16 times higher than the parent.

Terramycin resistance in *Bacillus megaterium* was studied by Northrop[19] using a
turbidostat. By growing sensitive populations at various concentrations of terramycin,
he was able to select for mutants at specific mutation rates. These differences in rate
could be accounted for by the decrease in growth rate of wild type under the stressful

situation. Evidently, the sensitive strain grew more rapidly than the resistant mutant at 0 or 0.1 γ/ml of the antibiotic; at higher concentrations the resistant populations were heavily selected for.

Resistance to bacteriophage has also been a focal point for microbiologists interested in continuous culture. Cocito and Bryson[20] found that a T_3 resistant mutant of *E. coli* B (a proline-requiring strain) displaced the sensitive parent from a nutrient broth turbidostat. Evidently, the parent's growth was being repressed by a protein inhibitor secreted into the growth medium. The mutant T_3^R was, however, no longer sensitive to the inhibitor and was therefore able to grow more rapidly than the parent in the environment of the turbidostat. The inhibitor turned out to be a colicine.[21]

An unusual observation has been reported by Taylor and Sleytr.[23] They observed that their *E. coli* strain was able to grow in chemostat under carbon-limitation over a wide variety of dilution rates. Attempts at establishing a steady state under magnesium-limitation resulted in "wash-out" of the culture. In fact, lysis was being observed with release of viral particles. Since cyclic AMP has been implicated in many repression-derepression systems,[24] it is possible that cAMP may control this response. Thus magnesium limitation may result in the suppression of adenylate cylase activity or synthesis, and consequently result in lowered cAMP levels. There is no evidence in the literature to support or refute this theory.

Continuous cultures have also been used to study mutation processes on many occasions. Ordinarily mutation rates are expressed as mutants produced per parent cell per generation. This observation of the dependence of mutation rate on generation and not time *per se* was responsible for the "copy-error hypothesis of mutation". In that theory, mutants arise because of errors in DNA replication. Since DNA replication is coordinately linked to cell division under all conditions, mutation rate is proportional to the rate of cell division or growth. Novick and Szilard[3] however, found that in tryptophan-limited chemostat cultures of the trp$^-$ *E. coli* B/1 and B/r/1, the rates of spontaneous mutation or of caffeine-induced mutation to T_5^R were independent of growth rate, but were proportional to time. These paradoxical observations were confirmed by other workers (reviewed by Kubitschek[25]) for methionine- and tryptophan-limited cultures of other strains.

These observations were not easily explained. Glucose-limited cultures, however, did show mutation rates, induced spontaneously and by caffeine, dependent on growth rate and not time.[26,27] This was shown to be true for lactate, succinate, or phosphorus-limited cultures. Clearly, the situation of amino acid-limitation could perhaps lead to different types of mutational events, one event apparently occurring continuously while the other occurring during DNA replication only.

Bendigkeit[28] resolved the situation by observing that glucose-limited cells contained only one nucleus, while tryptophan-limited cells were multinucleate. The number of genomes or nuclei was proportional to generation time. Thus cells growing at a doubling time of 2 hr contain one nucleus, while those growing at a doubling time 8 hr contain four. Thus their apparent replication rates of the cell or multinucleate structures were identical.[25] Consequently, mutation rates in both glucose and tryptophan-limited cultures were dependent on real growth rate.

The effectiveness of mutagenic agents or treatments can be ascertained in continuous culture easily. Routinely, a chemostat is set up of the required organism to be tested. The spontaneous mutation frequency can be easily calculated by a suitable screening procedure. Then the cells growing in culture can be exposed to the agent or treatment. The number of mutations appearing can then be ascertained (Figure 5). This has been used successfully by several workers. An excellent review by Kubitschek[25] covers the work on several mutagens, notably caffeine, 2-aminopurine, acridine orange and visible light, visible light alone, near U/V, far U/V and other agents.

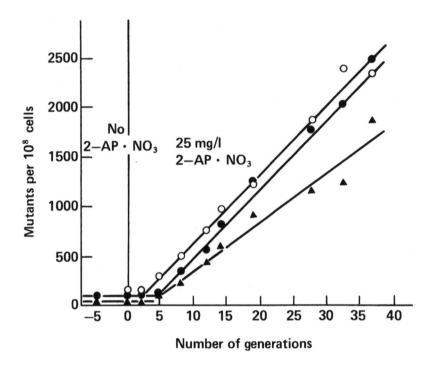

FIGURE 5.   Accumulation of mutants induced with 2-aminopurine in tryptophan-limited cultures of *E. coli* B$_d$1, trp. Generation time, 5.1 hr; expressed T5-resistant mutants; latent plus expressed T5-resistant mutants; expressed T6-resistant mutants. (From Kubitschek, H. E. and Bendigkeit, H. E., *Mut. Res.*, 1, 208, 1964. With permission.)

Kubitschek and Bendigkeit[26] predict that mutagenic treatments could be divided into two types. One type would act at the replication fork of DNA only while replication was progressing. Therefore the rate of mutagenic action of these compounds or treatments would be proportional to the replication rate of the DNA and therefore generation time of the culture. The other types of treatments or agents act after synthesis. These treatments would then yield mutation rates independent of generation times, however, they would be proportional to concentration of agent.

To test these two ideas, they set up chemostat cultures of *E. coli* limited by glucose. Under these conditions, no repair of damaged DNA was evident. In addition, the number of genomes per cell was independent of growth rate. They then examined two known mutagenic treatments, 2-aminopurine and U/V light. The former would be expected to act on replicating DNA, while the latter on DNA as a whole. As predicted, 2-aminopurine induced mutations accumulated at rates that were proportional to generation rate, while U/V induced mutations accumulated at rates independent of generation rate.[27]

Some very effective mutagens for batch culture (such as triethylene melamine and manganous chloride) are without much effect on continuous cultures of bacteria.[25] The reason for this variation has not been explained. Caffeine, an efficient mutagen, appears to act on continuous cultures of glucose-limited *E. coli* with kinetics that support the action of the compound on replicating DNA. Apparently, caffeine is bound to the bacteria during aerobic growth. Its binding also appears to follow saturation type kinetics and is dependent on DNA synthesis and energy production.[25] Further evidence indicates that the caffeine-DNA bond is formed when DNA is replicating, and that this binding is reversible by certain antimutagens. Consequently, errors in the

DNA may very well be induced on binding of caffeine during replication by alterations in the recognition code for the incorporation of the nucleotides.[28]

As discussed before, mutation rates induced by 2-aminopurine were dependent on generation rates, which implicates the replicating fork of DNA as the site of reaction.

Other treatments such as acridine orange in the presence of visible light, visible light and near and far U/V light give mutation rates independent of growth rate. This indicates that the sites of action are the DNA molecules in general, and not the replicating fork specifically.

Clearly, one advantage of continuous culture is that reproducible populations can be set up whose physiological properties are constant with respect to time. In addition, growth rates can be varied at will by the operator so that steady-state populations can be produced whose physiology is governed by growth rate and not by alterations in medium composition, temperature, and other factors. These factors, which are routinely used by microbiologists to alter growth rate, also alter physiological responses.[1] It is possible that physiological variation in bacterial population can alter mutation rates even in a constant genetic environment, although no work is available to validate this idea.

Antimutagens are compounds or treatments that decrease the rate of either spontaneous or induced mutagenesis.[29] The concentrations of these compounds are very often much lower than that used for mutagenic agents, so a direct detoxication is not involved. Since antimutagens are frequently toxic at high concentrations, continuous cultures are often employed to study the effect of these compounds.

Interestingly, antimutagens show specificity towards mutagenic agents. For example, the antimutagens adenosine and guanosine are effective against mutations induced by theophylline and caffeine, but totally ineffective against those produced by U/V or γ rays.[25]

Acridines (in the dark only) and polyamines are effective antimutagens against spontaneous, caffeine-induced, and 2-aminopurine induced mutation in glucose-limited cultures of *E. coli*.[25]

The study of repair of DNA damage by various agents in continuous culture has produced some very interesting results. The author does not intend to go into detail of the mechanisms involved in repairing lesions in DNA. They have been reviewed extensively by many workers.[30]

It is apparent that lethal lesions in the DNA of *E. coli* B/r can be repaired by dark repair mechanisms in glucose-limited chemostat cultures. The enzymes involved appear to be as efficient as nutrient broth batch culture grown cells, although the chemostat cells, grown at a growth rate tenfold less, are smaller, and have less DNA, RNA, and protein. Survival curves for both cells are identical.[25]

However, dark repair of mutational lesions in glucose-limited chemostat cultures is absent.[25] This type of repair is present in batch culture cells. At low doses of such agents as acridine orange, methylene blue (both agents in the light), and 2-aminopurine, all mutations both immediately expressed and latent, were expressed and remained constant after exposure.[25] This indicates that no repair of mutation lesions was occurring. Consequently, this would imply that the two repair processes, that of lethal and mutational lesions, must be operated by different mechanisms.[25]

In the preceding situations described, continuous chemostats were used. With these experiments, very few cells were killed by the mutagenic treatment. Another approach to study repair has employed discontinuous chemostat cultures.[25] In these studies glucose-limited chemostats were set up, and prior to administration of the mutagenic agent, the nutrient supply was halted. The dose of agent used was higher than that used in continuous chemostat cultures, and routinely more than 25% of the cells died.

After exposure to the agent, the medium supply was started at the required rate. Under these situations, not all the latent mutants were finally expressed indicating a repair process was involved. Repair was more prominent under slow growth conditions than fast. This indicated a competition between replication and repair for the potential mutant site. If repair occurs first, then wild-type progeny was produced. If replication occurred first, then one daughter cell would be mutant while the other would be wild-type.

They[25] proposed that repair of mutational lesion was absent in chemostat because repair required more glucose than ordinarily present in glucose-limited chemostats. However, in discontinuous chemostats, the killing of a large proportion of the population left more glucose for the survivors, which then promoted repair. Simply adding a "spike" of glucose to continuous glucose-limited chemostats after low doses of U/V did not, however, initiate repair.[25] These workers postulated that the repair system for mutations is activated or induced by high doses of U/V, requires departure from steady state, and requires and increased energy supply. These experiments were only performed with *E. coli* using the mutational event of transition between sensitivity and resistance to the phage T_5. Clearly, to see whether this is a general phenomenon or not, other loci need to be tested. Continuous cultures appear to allow very careful control of the enzymes for repair, and should prove useful in the future.

Continuous culture also offers a very valuable tool for studying evolution. Evolution occurs in chemostats when a particular cell type is selected over another, and may be specific or nonspecific in nature. Specific selection is involved when selection is operating upon a particular mutant under study. Nonspecific selection occurs when the frequency of a mutant is altered by selection for an unrelated mutation. Figure 6 shows a hypothetical situation depicting the selection of resistant cells in a chemostat although the population has not been exposed to the agent. As time progresses, the level of resistant cells increases, due to a spontaneous mutation event. Since the sensitive and resistant cells have the same growth rate, there is no selection. The level of resistant cells in the population will increase linearly until a steady state is attained. This is an example of nonspecific selection. However, before that time, the level of resistant cells in the population will decrease back to the threshold, since a new cell line will be selected for with "better" growth properties for the chemostat. These "better" adapted cells will grow more rapidly than the "old" cells under the conditions established in the chemostat. Since it is more likely that the "new" cell will be sensitive (the level of resistance would be very low in the population), the new chemostat population will become sensitive. This is an example of specific selection. This type of transition has been studied by Novick.[22] He showed that the new fitter strain of *E. coli* selected in a tryptophan-limited chemostat had a lower K_t for tryptophan than the original. Thus the new "fitter" strain would grow more rapidly than the parent under these conditions. In chemostats there is a continuous production of mutations that theoretically could produce mutant strains. Evolution involves the specific selection of those mutant strains — capable of higher growth rates under the conditions defined in the reaction vessel — than the previous strain. Nonspecific mutations occur along the time course of evolution, but are usually *cul de sacs* that lead nowhere evolutionary. Figure 7 gives a diagrammatic model of how selection and evolution appear to progress.

Evolutionary changes that can occur in continuous culture do not always result in an increase in affinity for a substrate for the new type. An example of evolution in chemostat is the selection for strains of *E. coli* and *K. aerogenes* that produce hyper levels of β-galactosidase under lactose-limited conditions. This has been dealt with in more detail earlier in this chapter.

Other examples of evolution are described by Hader et al.[31] Often, loss of biosyn-

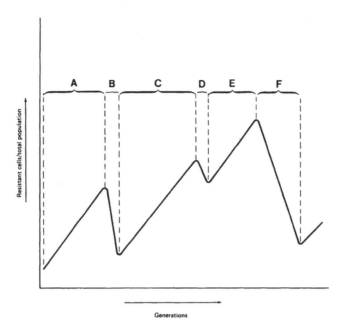

FIGURE 6. Hypothetical variation in levels of resistant cells in continuous culture of bacteria. At a time 0, a certain level of resistant cells (to antibiotic or bacteriophage) are present in the population. As growth progresses, the apparent forward mutation rate (sensitive → resistant) is greater than the backward. Hence the level of resistant cells increases in the population (A). However both sensitive and resistant cells are equally adapted for growth in the continuous culture and one is not selected over the other. With time a steady state could be attained where the level of resistants in the population is independent of time and the forward and backward mutation rate are equal. However more likely another mutation occurs in the population. This mutation makes the cell better equipped for growth in the conditions prevailing in the continuous culture. Since there are more sensitive cells than resistant in the population, then it is likely that the new, better adapted mutant is also sensitive. Thus all three cell types will compete in the continuous culture. The newly adapted, sensitive cell will out compete the others. As this cell type takes over the continuous culture, the level of resistant cells decreases (B). The process can repeat itself *ad infinitum* (C → F).

thetic function can be of a selective advantage under nonstressing situations. For example in histidine sufficient medium, his⁻ *B. subtilis* is selected over his⁺ and takes over the culture.[32] The phenomenon appears to be general, since an indole-requiring mutant outgrew an indole-producing strain in indole-sufficient medium. If the lesion is early in the enzyme pathway, then more advantage will be conferred to that strain over one with a late lesion. Thus an anthranilate-requiring strain (early block) of the organism outgrew a tryptophan-requiring (late block).[32] Apparently the energy saved is greater when the block is earlier. This lower energy load incidently allows the cell to redirect energy for growth and thus have a selective advantage. Metabolic economy thus may select not only for point mutants that possess lower levels of the nonessential enzymes, but for deletions that no longer synthesize DNA, mRNA, or even protein corresponding to the deleted gene. Such concepts easily predict the evolution of strains carrying multiple deletions for "nonessential" functions.

Another interesting example of selection in a chemostat involves an evolutionary

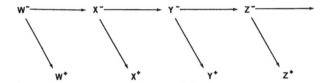

FIGURE 7.   Hypothetical flow chart illustrating the time course of
evolution in a chemostat. At time 0, the population contains W⁻ cells
only. As time progresses these mutate to W⁺; A very rare event (compared to the W⁻ → W⁺ transition) occurs causing W⁻ to X⁻ transition.
X⁻ is better adapted to the environment than W⁻ so the former is selected. In addition X⁺ disappears since it is at a selective disadvantage
compared to X⁻. Also shown are the other transitions eg Y⁻ → Y⁺ and
Y⁻ → Z⁻. The nonspecific selections, eg W⁻ → W⁺ are *cul de sacs* and
evolutionarily do not "lead anywhere". The W⁻ → X⁻ is a main stream
selection that leads to a "better adapted" cell line at least under the
conditions of growth in the chemostat.

important characteristic. It has been recognized for some time that some strains of
bacteria show a higher or lower rate of mutation than others. In these situations, the
characteristic was hereditary and the gene responsible could be mapped. Some situations indicated that the gene responsible for lower mutation rates mapped at DNA
repair loci such as exr in *E. coli* B and lex in *E. coli* K12.[29] Similarly, rec A⁻ strains
showed a much lower rate of mutation in the survivors of U/V light treatment.[29] In
these cases, it is predicted that the survivors of U/V treatment are unable to reapir
their DNA by an error prone mechanism and hence unable to introduce mutations into
the DNA. Without repair the cell dies.[29] On the other hand, Zamenhof[33] has examined
another gene of *E. coli*, ast, that somehow can cause bidirectional transition mutagen-
sis. The precise mechanism is not understood. In a similar fashion, Johnson and
Bach[34,35] have characterized a gene (Treffer's mutator gene) that is known to cause
AT→GC transition. Then two types of genes are recognized; those that cause an overall
increase in mutagenesis and are termed mutators, and those that cause a decrease and
are called antimutators. Muzyczka et al.[36] and Hershfield and Nossal[37] have presented
evidence to indicate that antimutators code for changes in DNA polymerase, so that
on replication of DNA errors are introduced.

As pointed out by Harder et al.,[31] an increased mutation rate would increase the
rate of change of a population and give it a selective advantage over slower mutating
populations. But since unfavorable and favorable mutations should still be produced
in similar proportions, the increased mutation rate would only be an advantage to the
population if mutation was the limiting factor. Too high a rate could, however, result
in a poor conversion of nutrients to "adapted" biomass, which would be analogous
to a decreased growth rate for the "well-adapted" cell.

The question of the evolutionary advantage of an increased mutation rate has been
posed and approached by several groups, namely Cox and Gibson[38] and Nestman and
Hill.[39] In both these studies continuous cultures were employed.

Cox and Gibson[38] examined the competition between two isogenic strains of *E. coli*,
one carrying the wild type allele mut⁺, and the other the mutant gene mut T₁. Under
glucose limitation the frequencies of mut T₁/mut⁺ increased as time progressed, indicating that the mutant had a selective advantage over the wild type. To differentiate
the strains they used the ability to utilize lactose. They concluded that the advantage
was conferred not by the intrinsic property of the mutator gene (contrast other selections discussed), but by the increased mutation rate associated with the gene. This

conclusion was supported by several findings, the final established mut T_1 mutant was different from the original mut T_1 and the mut$^+$ strains that were initially inoculated into the chemostats. Property changes that were detected included increased stickiness of the strain, such that it grew readily on the culture wall. This wall growth would allow selective retention of the cells in the culture vessel (see for comparision the work of Munson and Bridges[6] cited earlier in the chapter). Another property that was altered was the ability of the cell to survive glucose starvation. Clearly, this would have a selective advantage for the cells, since the level of glucose in the glucose-limited chemostat is very low. The second supporting line of evidence was that the mutations to stickiness and resistance to glucose starvation were independent of the presence of the mut T_1 mutation. Strains that lost the mut T_1 mutation did not lose the other properties. Clearly then, under certain circumstances, increased mutation rates are evolutionary advantages.

This review has attempted to demonstrate that while physiology and biochemistry of cells is constant in continuous culture, the careful monitoring of the properties of the cells is essential to demonstrate that genetic changes do not occur. If they do, then all is not lost; there are areas of genetics that can be studied in continuous culture. The physiologist or biochemist of today who does not beware may be the geneticist of tomorrow.

REFERENCES

1. **Tempest, D. W.**, The place of continuous culture in microbiological research, *Adv. Microb. Physiol.*, 4, 223, 1970.
2. **Novick, A. and Szilard, L.**, Description of the chemostat, *Science*, 112, 715, 1950.
3. **Novick, A. and Szilard, L.**, Experiments with the chemostat on spontaneous mutation of bacteria, *Proc. Natl. Acad. Sci. U.S.A.*, 36, 708, 1950.
4. **Strange, R. E. and Hunter, J. R.**, "Substrate-accelerated death" of nitrogen-limited bacteria, *J. Gen. Microbiol.*, 44, 255, 1966.
5. **Bryson, V.**, Microbial selection, *Science (N.Y.)*, 116, 48, 1952.
6. **Munson, R. J. and Bridges, B. A.**, "Take-over" — an unusual selection process in steady-state cultures of *Escherichia coli*, *J. Gen. Microbiol.*, 37, 411, 1964.
7. **Calcott, P. H.**, Nutritional factors affecting microbiol response to stress, *CRC Handbook of Nutrition and Food*, CRC Press, West Palm Beach, Fla., 1978.
8. **Novick, A.**, Bacteria with high levels of specific enzymes in growth in living systems, *Purdue Growth Symposium*, Basic Books, New York, 1961, 93.
9. **Smith, R. W. and Dean, A. C. R.**, β-galactosidase synthesis in *Klebsiella aerogenes* growing in continuous culture, *J. Gen. Microbiol.*, 72, 37, 1972.
10. **Calcott, P. H. and Postgate, J. R.**, The effects of β-galactosidase activity and cyclic AMP on lactose-accelerated death, *J. Gen. Microbiol.*, 85, 85, 1974.
11. **Horiuchi, T., Tomizawa, J., and Novick, A.**, Isolation and properties of bacteria capable of high rates of β-galactosidase synthesis, *Biochim Biophys. Acta*, 55, 152, 1962.
12. **Novick, A. and Horiuchi, T.**, Hyper-production of β-galactosidase by *Escherichi coli* bacteria, *Cold Spring Harbor Symp. Quant. Biol.*, 26, 239, 1961.
13. **Vojtisek, V., Sikyta, B., and Slezak, J.**, Regulation of the hyper-production of β-galactosidase in continuous culture of *Escherichia coli*, B, in *Continuous Cultivations of Microorganisms*, Malek, I., Beran, K., Fencl, Z., Munk, V., Ricica, J., and Smrckova, H., Eds., Academic Press, London, 1969.
14. **Von Hofsten, B.**, The inhibitory effect of galactosides on the growth of *Escherichia coli*, *Biochim Biophys Acta*, 48, 164, 1961.

15. **Clarke, P. H.**, The evolution of enzymes for utilization of novel substrates, in *Evolution in the Microbiol. World.*, 24th Symp. Soc. Gen. Microbiol., Cambridge University Press, London, 1974, 183.

16. **Hegeman, G. D.**, Synthesis of enzymes of the Mandelate pathway by *Pseudomonas putida* III. Isolation and properties of constitutive mutants, *J. Bacteriol.*, 91, 1161, 1966.

17. **Cocito, C. and Vogel, H. J.**, Heritable Lowering of an Enzyme Level and Enzyme Repressibility Observed upon Continuous Culture of *Escherichia coli* in the Presence of a Repressor, in Proc. 10th Int. Congr. Genetics, Montreal, Canada, 1968.

18. **Munson, R. J.**, Turbidostats, in *Methods in Microbiology*, Vol. 2, Norris, J. R. and Ribbons, D. W., Eds., Academic Press, London, 1970, 349.

19. **Northrop, J. H.**, The proportion of Streptomycin-resistant mutants in *B. megatherium* cultures growing in continuous culture, in *Continuous Cultivation of Microorganisms*, Malek, I., Beran, K., and Hospodka, J., Eds., Academic Press, London, 1964.

20. **Cocito, C. and Bryson, C.**, Properties of colicine from *E. coli* strain B, *Bacteriol. Proc.*, 38, 1958.

21. **Bryson, V.**, Applications of continuous cultures of microbial selection in *Recent Progress in Microbiology*, Tunevall, G., Ed., Charles C Thomas, Springfield, 1959, 371.

22. **Novick, A.**, Mutagens and antimutagens, *Brookhaven Symp. Biol.*, 8, 201, 1956.

23. **Taylor, P. W. and Sleytr, V. B.**, Release of a lysogenic bacteriophage from a smooth urinary *Escherichia coli* strain following magnesium limitation in a chemostat, *FEMS Microbiological Letters*, 2, 189, 1977.

24. **Perlman, R. L. and Pastan, I.**, Pleiotropic deficiency of carbohydrate utilization in an adenylate cyclase deficient mutant of *Escherichia coli*, *Biochem. Biophys. Res. Commun.*, 371, 151, 1969.

25. **Kubitscheck, H. E.**, Introduction to research with continuous cultures, *Prentice Hall Biological Techniques Series*, Prentice Hall, Englewood Cliffs, N.J., 1970.

26. **Kubitschek, H. E. and Bendigkeit, H. E.**, Mutation in continuous cultures. I. Dependence of mutational response upon growth limiting factors, *Mutat. Res.*, 1, 113, 1964.

27. **Kubitschek, H. E. and Bendigkeit, H. E.**, Mutation in continuous cultures. II. Mutations induced with ultra-violet light and 2-aminopurine, *Mutat. Res.*, 1, 208, 1964.

28. **Bendigkeit, H. E.**, Caffeine Induced Mutation in *Escherichia coli* Ph.D. thesis, Illinois Institute of Technology, 1968, quoted by H. E. Kubitschek, H. E., Introduction to research with continuous cultures, *Prentice Hall Biological Techniques Series*, Prentice Hall, Englewood Cliffs, N.J., 1970.

29. **Clarke, C. H. and Shankel, D. M.**, Antimutagenesis in microbial system, *Bacteriol. Rev.*, 39, 33, 1975.

30. **Bridges, B. A.**, Bacterial reaction to radiation, *Pattern of Progress in Microbiology*, PPM8. Meadowfield Press, Durham, England, 1976.

31. **Harder, W., Kuenen, J. G., and Matin, A.**, A review: microbial selection in continuous culture, *J. Appl. Bacteriol.*, 43, 1, 1977.

32. **Zamenhof, S. and Eichhorn, H. H.**, Study of microbial evolution through loss of biosynthetic functions: establishment of defective mutants, *Nature (London)*, 216, 456, 1967.

33. **Zamenhof, P. J.**, On the identity of two bacterial mutator genes: effects of antimutagens, *Mutat. Res.*, 7, 463, 1969.

34. **Johnson, H. G. and Bach, M. K.**, Apparent suppression of mutation rates in bacteria by spermine, *Nature (London)*, 208, 408, 1965.

35. **Johnson, H. G. and Bach, M. K.**, The antimutagenic action of polymines: suppression of the mutagenic action of an *E. coli* mutator gene and of 2-aminopurine, *Proc. Natl. Acad. Sci., U.S.A.*, 55, 1453, 1966.

36. **Muzyczka, N., Poland, R. L., and Bessman, M. J.**, Studies on the biochemical basis of spontaneous mutation. 1. A comparison of the DNA polymerases of mutator, antimutator, and wild type of bacteriophage T₄, *J. Biol. Chem.*, 247, 7116, 1972.

37. **Hershfield, M. S. and Nossal, N. G.**, *In vitro* characterization of a mutator T₄ DNA polymerase, *Genetics*, 73, 131, 1973.

38. **Cox, E. C. and Gibson, T. C.**, Selection for high mutation rates in chemostats, *Genetics*, 77, 169, 1974.

39. **Nestman, E. R., and Hill, F. R.**, Population changes in continuously growing mutator cultures of *Escherichia coli*, *Genetics*, 73, 41, 1973.

40. **Lee, S. K. and Calcott, P. H.**, unpublished observations.

Chapter 6

CONTINUOUS CULTURE OF PLANT CELLS*·**

W. G. W. Kurz and F. Constabel

* This chapter was submitted December 1977.
** NRCC No. 16285.

I. INTRODUCTION

Plant cells cultured in vitro have come into prominence, because of their potential for unlimited growth and for plant regeneration. This includes the synthesis and accumulation of secondary metabolites. Applied to economic plants the method of culturing cells in vitro may significantly improve plant propagation, plant breeding, and the industrial production of physiologically active substances.[1] The large scale propagation of various plants[2] and biotransformation of steroids and cardiac glycosides[3,4] by means of cultured plant cells demonstrates the relevance of the cell culture method to agriculture and industry.

Plant cell culture is defined as a method of growing cells derived from seed plants under controlled environmental conditions in vitro. The term suspension culture refers to cells and cell aggregates growing dispersed in an agitated liquid medium. When compared with microbial cultures, cell suspension cultures can be characterized as follows:

1.   In addition to mineral salts and carbohydrates they require phytohormones for growth.
2.   The cell generation time generally exceeds 24 hr.
3.   The cells are diploid, occasionally poly- and aneuploid.
4.   They possess a morphogenetic potential.

The objectives of employing plant cell cultures are directed towards morphogenesis, i.e., the analysis and control of growth and differentiation, as well as metabolite accumulation in cells. So far, research activity has resulted in a wealth of information on the effect of externally applied morphogenetic factors: nutrients, phytohormones, light, and temperature. Morphogenesis-inducing principles remain to be elucidated. The establishment of open continuous cultures, in particular synchronous cultures greatly enhances possibilities for better understanding induction processes. In amplifying cellular events, synchronous cell cultures may permit the detection of early signs of morphogenetic reactions on a molecular level. Continuous cultures could also render industrial exploitation of plant cell cultures more economical.

II. ESTABLISHMENT OF CELL SUSPENSION CULTURES

The inoculum for continuous cultures generally are batch cultures of cell suspensions. The establishment of such cultures has been described in great detail, and two handbooks provide step by step procedures.[5,6]

Plant cell cultures are initiated by excising sections of parenchyma from a given plant and transferring these explants to solidified nutrient media. Seedlings often are found to be the most suitable source. Within a month these explants should produce callus growth. Testing a variety of media will enhance chances for callus formation. The media proposed by Murashige and Skoog[7] and by Gamborg et al.[8] supplemented with 1 mg l^{-1} auxin [2,4-dichlorophenoxyacetic acid (2,4-D) or 1-naphthaleneacetic acid (NAA)] are recommended.[9] Once new growth around explants has yielded a sizable layer of cells, it is isolated and serially subcultured. The texture of the callus may vary from soft to friable and hard. Varying the concentration and combination of phytohormones may significantly affect the texture. Soft and friable callus can be subcultured in liquid medium and would render cell suspension cultures without difficulties.

In general, cells are cultured as 50 or 100 ml suspensions in 250 or 500 ml Delong flasks on gyratory shakers (100 to 150 r/min) at 25 to 28°C in day and night cycles or continuous light of less than 2000 lux. Rapidly growing cultures will require transfers of 10 m*l* suspension to 50 m*l* fresh medium every 2 to 3 days.[9] More complex media may permit more dilute inocula and considerably longer culture periods.[10]

Cell growth has been studied extensively with a few well-established cultures derived from *Acer pseudoplatanus,*[11,12] *Glycine max,*[13] *Haplopappus gracilis,*[14] *Nicotiana tabacum,*[15] and *Ipomoea* sp.[16] Typically the growth of these cells follows a sigmoid curve.

Cytological examination of a cell suspension culture may show a lack of uniformity. The occurrence of cells that have acquired special structures (elongated cells, tracheids) or have accumulated secondary metabolites (pigments) would attest to the presence of mature cells that have temporarily or definitely been sequestered from cell cycles.[17] Modifications in nutrients resulting in more rapid growth may eliminate differentiation and enhance uniformity in cell suspension cultures. A special class of cells has been described for a soybean cell line. These Q-cells fail to divide or synthesize DNA, but do synthesize RNA and protein, although at a greatly reduced rate.[18]

Cell lines that are known for their potential for the synthesis of secondary metabolites are *Rosa* spec.,[19] *Glycine max,*[20] *Petroselinum hortense,*[21] *Haplopappus gracilis,*[22] *Daucus carota,*[23] *Catharanthus roseus,*[24] *Digitalis lanata,*[25,26] *Morinda citrifolia,*[27] and *Ruta graveolens.*[28] Cell lines may lose their capacity for metabolite production during long term culture.[17]

III. GENETIC STABILITY OF PLANT CELL CULTURES

The genetic stability of cell suspension cultures is of great concern. Chromosome and gene mutations in cell cultures are well documented.[29] Although there is evidence that polyploid cell lines may arise from endo-reduplicated nuclei of the explant, the variability of chromosome number and structure observed in established cultures strongly indicates that changes result from culturing. *Vicia hajastana* cultures, for instance, initially had a very low frequency of diploid cells. After a period of 12 months the frequency had increased to 91% indicating a strong selection for diploid cells.[30] A carrot culture analyzed at various times after initiation showed stable chromosome numbers and karyotypes with small, but significant variation about the modes. Some polyploid multiples of the model chromosomes were present in all carrot cell lines at low frequency.[31] As in *Vicia,* competitive selection of diploid cells may have prevented the culture from continuously producing cells with polyploid, aneuploid, and structurally altered karyotypes.

It appears that even under most favorable culture conditions, cytological deviations are constantly arising and are subject to selection pressure predominantly for the capacity to divide. Any selective advantage in this respect will cause a drift and temporary heterogeneity towards a population with an altered chromosome mode of relative stability.

Duration of growth periods between transfers to fresh media and composition of the phytohormone supplement appear as means to regulate the cytological constitution of cultured cells, and may enhance the ability to maintain genetically defined plant cell cultures. For maintenance of cell lines characterized by a special biosynthetic potential deep-freeze preservation may be considered.[32]

IV. CULTURE UNITS

A serious disadvantage of most plant cell lines grown as suspension culture is the strong tendency for the cells to grow as cell aggregates, resulting in heterogeneous cell populations. These types of populations may be adequate for production of biomass — or where plant cells are employed in the production of secondary metabolites, but are unsatisfactory for many metabolic and physiological studies, where the use of more uniform cell material grown under defined conditions in a state of balanced growth is desirable.

Conventional microbial fermentor systems may in specific cases be successfully adapted to the cultivation of plant suspension cultures, however, because of shearing effects and other mechanical features that would rupture the thin-walled plant cell, they are often unsuitable.

All culture units are designed to produce cells that are evenly distributed within the culture, and where thorough mixing of medium and cells and an adequate gas exchange between the liquid and gas phase is provided. In the past few years several culture units were built that fulfill these requirements, and can be used to grow uniform cell material under defined conditions in continuous culture.

In the culture unit described by Kurz (Figure 1),[33] regular pulses of compressed air provide both agitation and aeration by feeding pulses of compressed air at regular intervals into the culture vessel, a flat bottomed glass cylinder. As the compressed air (5 to 10 psi) enters the cylinder via a central port at the base, it expands into a large bubble having the same diameter as the culture vessel. As the bubble moves upward through the cylinder, the entire culture passes as a thin layer between the wall of the vessel and the surface of the air bubble, thus being effectively aerated and mixed. Furthermore the reduction of air pressure and expansion of the bubble causes vibration in the culture, which probably is the main factor in producing a culture predominantly consisting of very small aggregates and single cells. When operated as a chemostat, the unit is fitted with a dual pump both supplying the culture with a constant flow of fresh medium, as well as withdrawing an equal amount of cell suspension. This culture unit has successfully been employed in the induction of partial synchrony of cell division in continuously grown plant cell suspension cultures.[34,35]

A V-shaped culture unit was developed by Veliky and Martin (Figure 2).[36] In their system, the culture is agitated by a teflon-coated double bar magnet stirrer supported on a short glass rod at the base of the vessel. The air is supplied through a hypodermic needle. The buildup of cell clusters is minimized by reducing the number of devices inserted into the culture. This culture unit is mainly used for semicontinuous culture through intermittent renewal of medium and harvesting of culture, as well as in batch operations.

The Phytostat described by Miller and co-workers (Figure 3),[37] as well as a similar unit by Wilson and co-workers[38] may be used for continuous and batch operations. In both units the culture vessel is a round-bottom flask, fitted with a magnetic stirrer and sintered glass aerator. They are both equipped with needle valves for the automatic collection of samples. However, because stirrer, aerator, and similar devices protrude into the culture, the buildup of cell material on these at the interface between liquid and gas phases poses a problem.

The culture units described above have been specifically designed to grow plant cell suspensions in batch or continuous culture over long periods of time necessitated by the longer generation times of plant cells. In designing these units special consideration

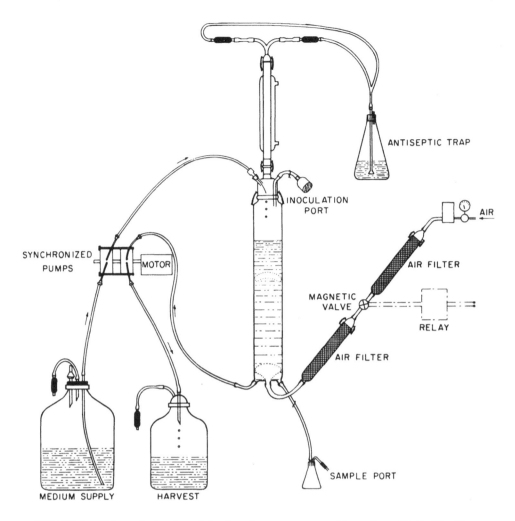

FIGURE 1. Schematic drawing of the Kurz fermentor system. (From Kurz, W. G. W., *Tissue Culture: Methods and Applications,* Kruse, P. F., Jr. and Patterson, M. K., Eds., Academic Press, New York, 1973, 359. With permission.)

has been given to the tenderness of the cell material and any mechanical features that may lead to damage of the thin-walled plant cells has been avoided.*

V. CONTINUOUS CULTURE SYSTEMS

The ideal state for a cell suspension culture is one of morphological, biochemical, and genetic homogeneity when grown in a fully controlled environment. A cell derived from a seed plant differs from a microbe in its requirements, and the culture techniques developed for the latter are not directly adaptable. Plant cell suspension cultures are generally heterogeneous with both single cells and aggregated cells growing at different rates. Their proportion and size depend on species, age of culture, composition of medium, and physical environment. Considering all these aspects it is not surprising that numerous attempts to achieve ideal conditions for cell division and uniformity have only partly been successful.

* Since completion of the manuscript, multiliter bioreactors based on the principle of agitation and aeration by means of airlift have been introduced.[53]

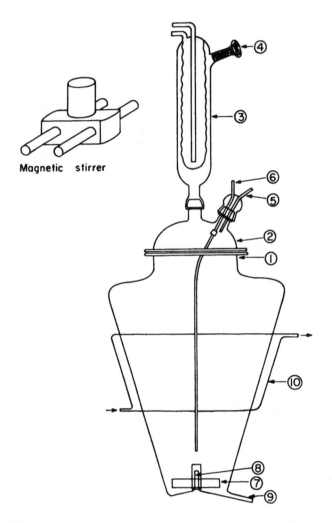

Magnetic   stirrer

FIGURE 2.   Schematic drawing of the V-Fermenter: 1. Large-diameter flat-flange joint. 2. Multisocket lid. 3. Water-cooled condenser. 4. Air exhaust with a sterilizing filter. 5. Inlet for medium. 6. Air inlet. 7. Teflon®-coated double-bar magnetic stirrer. 8. Short glass rod. 9. Sampling outlet. 10. Water jacket. (From Veliky, I. A. and Martin, S. M., *Can. J. Microbiol.*, 16, 223, 1970. With permission.)

Several semicontinuous and continuous culture systems for plant suspension cultures have been developed in the past few years to satisfy the need of more uniform cell material for metabolic and physiological studies.

A. Semicontinuous Culture Systems

In this type of culture, part of the culture volume is removed and at infrequent intervals replaced by an equivalent amount of fresh medium. Semicontinuous culture has been mainly employed for maintaining stock cultures in an exponential growth phase over longer periods of time by avoiding nutrient limitation through frequent subculture. Such cultures serve as a ready source of uniform inocula,[36] or are used in metabolic and physiological studies.[39-43] However, data[36,44] obtained indicate that balanced growth cannot be achieved by this cultivation method (Figure 4).

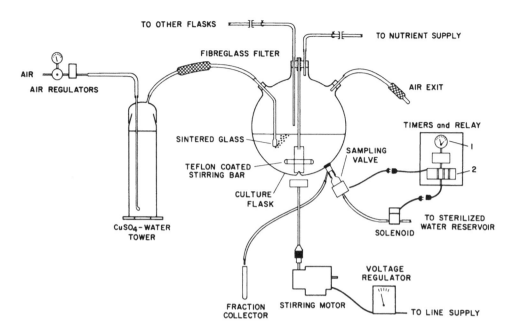

FIGURE 3.   Schematic drawing of the Phytostat. (From Miller, R. A., Shyluk, J. P., Gamborg, O. L., and Kirkpatrick, J. W., *Science (N.Y.)*, 159, 540, 1968. With permission.)

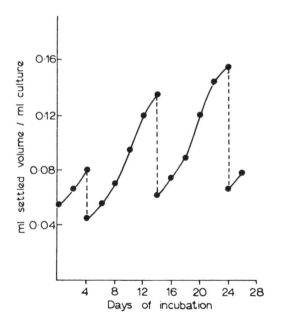

FIGURE 4.   Growth pattern of *Arachis hypogaea* grown in suspension culture under semi-continuous conditions. (From Verma, D. P. S. and van Huystee, R. B., *Exp. Cell Res.*, 69, 402, 1971. With permission.)

B. Closed Continuous Culture System

In a closed continuous system, the cells of the culture are retained in the fermentor and the inflow of fresh medium is balanced by withdrawal of an equivalent amount of spent medium, thus the number of cells increases continuously while the volume of the culture is kept constant. However the nutritional demand of such an exponentially growing culture can be met for a limited time only, and the availability of oxygen is eventually the limiting factor to growth. Wilson et al.[38] and Street et al.[45] described a system featuring a wide stilling tube associated with a siphon. The culture moves up the stilling tube and (due to the slow flow rate and gravity) cells and spent medium separate and only the latter flows over the apex of the siphon. The closed continuous system could be employed to study the effect of nutrients or metabolic precursors on the culture or the production of metabolites. Such a system should also allow the extracellular product to be continuously harvested.[46] However, it must still be determined how long cells can be maintained under these conditions.

C. Open Continuous Culture System

In open continuous culture, a continuous feed of medium flowing into a constant volume of growing culture and a constant withdrawal of an equal amount of culture produces a steady-state condition for growth. The environment imposes a constant growth rate on the cells so that the doubling time and overall metabolism of the cells remain constant and thus characteristics of the steady state. It is possible to change the growth rate by altering the flow through the system; in this way the growth of cells may be examined at different growth rates by analysis of the appropriate steady states and metabolic regulation in the production of metabolites is possible.

The open continuous system may be operated either as "chemostat" in which growth rate and cell density remain constant due to a fixed input rate of a growth-limiting nutrient in the medium; or as "turbidostat" where the cell density is set at a predetermined level and fresh medium added intermittently to maintain this density.

1. Chemostat

The chemostat system has several advantages over others in the investigation of cell metabolism, because of the possibility to maintain steady states of cell growth over long periods of time, and by controlling these through selection and concentration of a single growth-limiting nutrient. Furthermore this system not only allows investigation at the stage of balanced growth, but also during the period of unbalanced growth in the transition from one steady state to another when metabolic changes occur.

Within the biological limits of a cell line, any dilution made will result in a steady-state population with a specific cell density, which in turn is determined by the concentration of a growth limiting nutrient in the inflowing substrate, oxygen tension, pH, and other parameters. At each steady state and specific growth rate, a nutrient equilibrium is reached that can be expressed by the equation:

$$\frac{ds}{dt} = DS_R - Ds - \frac{\mu x}{Y} = 0$$

Where s = Equilibrium concentration of the limiting nutrient in the culture

S_R = Concentration of limiting nutrient in the inflowing substrate

μ = Rate of increase in biomass of the culture per unit biomass concentration (specific growth rate)

x = Cells per unit culture volume at time t (cell density)

D = Number of culture volumes replaced by inflowing substrate per unit time (dilution rate)

Y = Yield coefficient = amount of cells produced per unit of growth limiting nutrient consumed

The specific growth rate is determined by the equilibrium level of the limiting nutrient in the culture, which can be expressed as

$$\mu = \mu \max \frac{S}{K_s + S}$$

where K_s is a saturation constant that is numerically equal to the concentration of the growth limiting nutrient at which the specific growth rate is half of its maximum value. A steady state can therefore only be obtained if the specific growth rate does not exceed μmax. At dilution rates where μ would exceed μmax, growth becomes unbalanced and the cells will be gradually washed out of the fermentor.

Growth kinetic studies of plant cells in chemostats are few, but results indicate similarity to the ones obtained with microbes. However, because of the longer generation times, the heterogeneous populations, and lower metabolic rates, chemostat cultures of plant cells are more complex and cannot be compared directly to microbial cultures.

The initial chemostat work was mainly concerned with biomass production. Miller et al.[37] and Constabel et al.[22] obtained steady states of growth in *Glycine max* and *Haplopappus gracilis* cultures in the Phytostat by adding fresh nutrient continuously to a 2ℓ culture and withdrawing 10 mℓ portions of culture automatically at 20 min intervals. The cultures yielded 1.34 mg dry weight mℓ$^{-1}$ at a generation time of 46 hr. Kurz[47] reported chemostat cutures of *Glycine max* and *Triticum monococcum* grown in a fermentor using forced aeration for stirring at generation times of 30 hr, and cell yields of 1.3 mg dry weight mℓ.$^{-1}$ The cell line most intensely studied under chemostat conditions is *Acer pseudoplatanus*. Wilson et al.[38] were able to maintain steady states at different growth rates for long periods of time. As Figure 5 indicates the steady state under nitrogen limitation established not only constant levels in cell number, packed cell volume, or cell dry weight, but also in the consistency of protein, DNA, RNA, glucose, nitrate, and phosphate levels, as well as oxygen tension and pH. Similar results were obtained under phosphorous limitation.[48] King and Street[49] showed that in N-limited chemostat cultures of *Acer pseudoplatanus,* an increase in dilution rate is followed by a progressively steeper decline in the steady-state cell density (Figure 6), which indicates not only the limiting effect of a single nutrient on the growth rate, but also points to a high K_s value, (i.e., a low affinity for the limiting nutrient). The K_s values for *Acer pseudoplatanus* under nitrate limitation were found to be 0.13 mM and phosphorus limitation 32 μM, which is considerably higher than the values obtained from bacterial cutures. The activities of enzymes of the pentose phosphate pathway and the Embden-Meyerhof-Parnas pathway in chemostat cultures of *Acer pseudoplatanus* showed that at lower growth rates the in vitro activity of phosphofructokinase (EMPP) was much higher than glucose-6-phosphate dehydrogenase (PPP), whereas at fast growth rates the response was reversed.[50] Such changes in the pathways of carbohydrate oxidation are also reflected in changes of carbon utilization. Carbohydrate consumption is maximal at an intermediate growth rate, whereas respiration is at a minimum at this stage.[49] It should also be noted that considerable changes in metabolism and cell growth may occur, when the different forms of limiting nutrient are used. Young[51] found that if urea alone is used instead of NO_3^- or a combination of both, the cell densities at specific growth rates were lower, but the cell dry weight doubled over those grown on a urea-NO_3^- combination. Nitrogen assimilating enzymes vary as well with changing growth rates. Transaminases increase while urease, nitrate reductase, and glutamate dehydrogenase decrease in activity as growth rate increases. Differences like these, which occur at different specific growth rates under controlled culture conditions, open a wide field of opportunities in employing plant cells for the production of desirable compounds. King[52] could show that 2,4-dichloro-

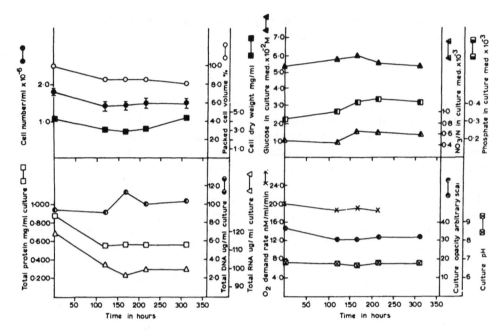

FIGURE 5.   A steady state established in a 4-liter chemostat culture of *Acer pseudoplatanus* cells. The culture was diluted for ca. 400 hr at a rate of 0.194 day⁻¹. Samples (50 mℓ) were withdrawn at intervals for biomass measurements, nutrient analysis, and respiration rate determinations. Culture opacity and pH were monitored continuously in the culture vessel. (From **Street, H. E., Ed.,** *Plant Tissue and Cell Culture,* University of California Press, Berkeley, 1973, 325. With permission.)

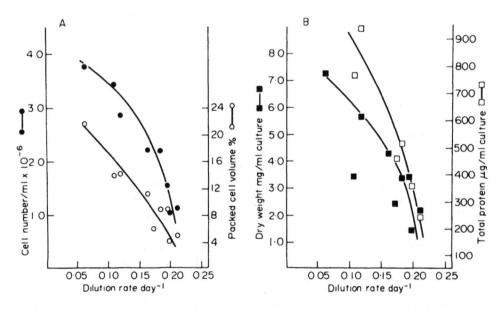

FIGURE 6.   Relationship between steady-state biomass and dilution rate in chemostat cultures of *Acer pseudoplatanus* cells. Each point in the curve for any one parameter in A and B represents the mean of a series of values obtained at intervals during a steady state established at the dilution rate indicated. The standard errors of these means were all less than 10% of the mean values. The data were obtained during nine different steady states (all in excess of 400 hr) established in five separate chemostat cultures. (From Street, H. E., Ed., *Plant Tissue and Cell Culture,* University of California Press, Berkeley, 1973, 327. With permission.)

phenoxyacetic acid (2,4-D) affects the level of production of phenolics dependent on specific growth rates.

Chemostat studies involving plant cells have only been carried out for the past few years and data acquired regarding metabolism and growth in this system are still limited. Nevertheless plant cells generally seem to follow the concept of the chemostat theory, and in the future this sytem will find its use in further studies of metabolic regulations and possibly the production of secondary metabolites.

2. Turbidostat

This system is preferable to the chemostat in cases where cultures are grown at low cell densities and the specific growth rate is close to or at μmax. As opposed to the chemostat (where the growth rate is determined by substrate limitation) the turbidostat is entirely regulated by nutritional and environmental conditions, as well as the metabolic capability of the cell itself. This makes the turbidostat system a valuable tool for investigating the regulatory mechanism of metabolism.

A turbidostat for plant cell suspension cultures has been described by Wilson et al.[38] In this system the cell density is monitored with the help of a density detector, which is linked to a control system regulating the input of fresh medium and culture volume and allows the population density to be kep at a predetermined value. Figure 7 shows steady states of a turbidostat culture of *Acer pseudoplatanus* at high growth rates. Eriksson[39] used turbidometric measurements to demonstrate a relationship between optical density and cell dry weight in cultures of *Haplopappus gracilis*.

D. Continuous Phased (Synchronous) System

In batch and continuous culture the cells divide randomly. In batch culture the doubling time of successive generations is likely to change, while in the continuous system (chemostat) this remains constant. In continuous phased growth, (an "open system") ideally all the cells are at the same stage of development in the cell cycle. Thus, instead of an average condition as is obtained in the steady state of the chemostat from the randomly dividing population of cells, in the continuous phased culture there is a pattern of change that coincides with the cell cycle and repeats itself with each successive doubling of the cell population. The cell, amplified by the size of the phased population may be examined at any stage of the cell cycle and at any desired growth rate. This has the advantage that enzymes or metabolites occurring only at certain stages in the cycle can be obtained at maximal yields.

This method of continuous synchronous cultivation of plant cells in suspension culture has been developed recently by imposing flushes of nitrogen or ethylene gas at regular intervals upon chemostat cultures at steady states.[34,35] Temporary anaerobic conditions (90 min/24 hr) did not adversely affect the cultures, and dry weight remained fairly constant over 5 weeks of treatment. The mitotic index of continuously aerated chemostat cultures was maintained at about 5%. In response to the treatment with N_2 for 90 min/day, the mitotic index decreased drastically (Figure 8). Within 8 hr after return to aeration, the cells resumed division with an increasing frequency and the mitotic activity followed 24 hr cycles with peaks well in advance of the N_2 treatment. When N_2 was omitted and the cultures were aerated without interruption, cell cycles of 24 hr continued for at least 4 days. When the intervals between N_2 treatments were lengthened to 30 hr, the populations reacted with 30 hr cycles and increased amplitudes (Figure 9). Partial synchrony was greatly improved by subjecting the cultures to a treatment of 3% ethylene for 3 hr, followed immediately by 3 hr of 3% CO_2 and 30 hr of aeration prior to the next ethylene treatment. This sequence of gases made the amplitude of cell cycles extend from MI (min) = 0.5 to 2% to MI (max) = 15 to 20% (Figure 10).

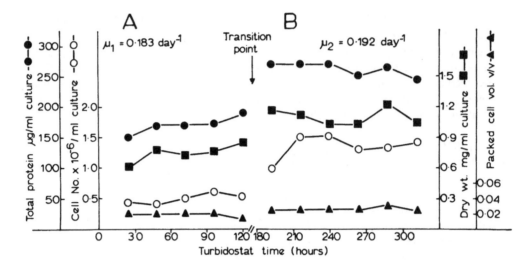

FIGURE 7.   Steady states at high growth rates in a turbidostat culture of *Acer pseudoplatanus.* After 120 hr operation in state A, the balance-point of a density-detection Wheatstone bridge circuit was raised by a potentiometer adjustment; fresh medium additions temporarily ceased. The culture density exceeded the new balance-point 48 hr later, and this initiated a further series of medium additions. A new steady state (B) rapidly stabilized at a growth rate (μ) not significantly different to that of state (A). (From Street, H. E., Ed., *Plant Tissue and Cell Culture,* University of California Press, Berkeley, 1973, 321. With permission.)

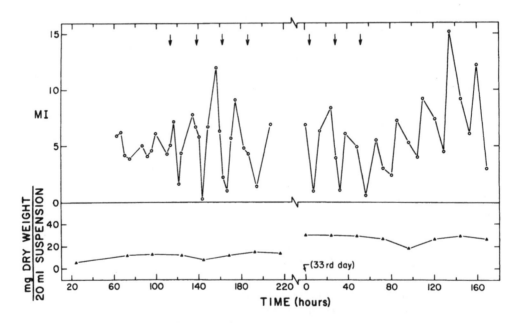

FIGURE 8.   Mitotic activity (MI%) and dry weight (mg/20 mϑ suspension) of 1.8ϑ soybean cell suspension cultured continuously and subjected to N_2 pulses through 90 min/ 24 hr. Arrows indicate beginning of N_2 treatments. (From Constabel, F., Kurz, W. G. W., Chatson, B., and Gamborg, O. L., *Exp. Cell. Res.,* 85, 105, 1974. With permission.)

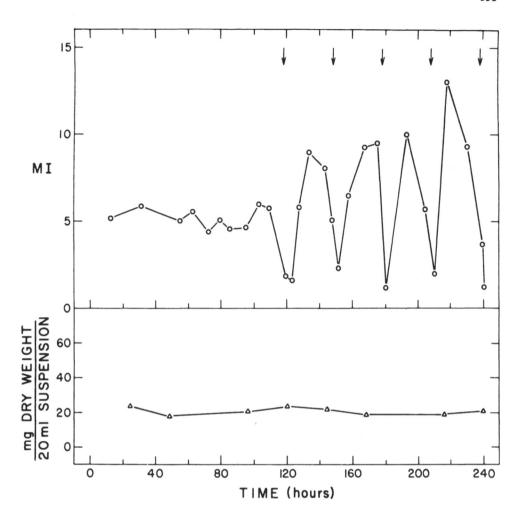

FIGURE 9.   Mitotic activity (MI%) and dry weight (mg/20 mℓ suspension) of 1.8ℓ soybean cell suspension cultured continuously and subjected to N₂ pulses through 90/30 hr. Arrows indicate beginning of N₂ treatments. (From Constabel, F., Kurz, W. G. W., Chatson, B., and Gamborg, O. L., *Exp. Cell Res.*, 85, 105, 1974. With permission.)

The availability of synchronized populations of plant cells greatly facilitates investigations of molecular and physiological events during the cell cycle and of the control mechanisms that govern mitosis. Furthermore, such populations would allow studies on cell competence for the uptake and integration of foreign genetic information, and on the compatability of nuclei in heterokaryons.

VI. CONCLUSION

Plant cells are amenable to long-term culture in continuous systems. The culture in these systems is restricted to cell material that is characterized by relatively rapid growth and that has a tendency to dissociate into small aggregates and single cells. Refinements in culture methods and expansion of the range of cell lines suitable for cultivation in continuous systems is essential for the future development of cell culture. The emphasis should be primarily on cell lines derived from plant species that have industrial value.

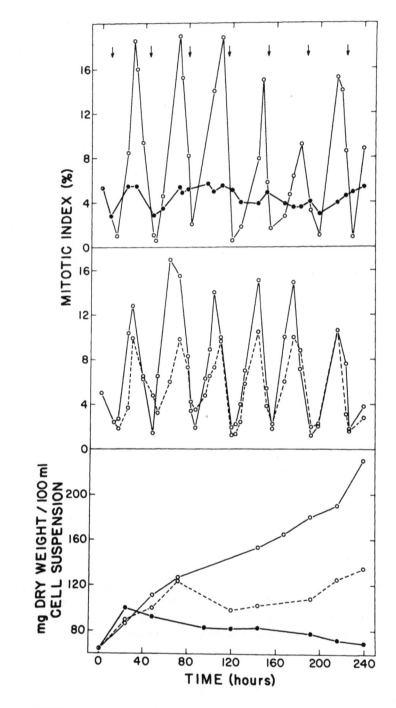

FIGURE 10.   Abscissa: time (hours); ordinate: (a,b) MI (%); (c) mg dry
weight/100 mℓ cell suspension.   Variation in the frequency of cell division
and growth of suspension cultures of soybean with time under a regime of
ethylene and CO_2 flushes (arrow) applied at 36 hr intervals. (a) ○, 3% ethylene 3 hr followed by 3% CO_2 3 hr; ●, control; (b) ○, 3% ethylene and
3% CO_2 3 hr. ○-----○, 3% ethylene 3 hr; (c) ○, 3% ethylene 3 hr followed
by 3% CO_2 3 hr; ○-----○ 3% ethylene 3 hr; ●, control. (From Constabel,
F., Kurz, W. G. W., Chatson, K. B., and Kirkpatrick J. W., *Exp. Cell Res.*,
105, 263, 1977. With permission.)

Secondary metabolites that are of ubiquitous occurrence and whose synthesis and accumulation do not require marked cytodifferentiation have already been demonstrated in various continuously cultured cell lines. In contrast, metabolites that are formed in highly differentiated (specialized) cells only, can hardly be expected to occur in continuous cultures, unless the necessary pattern of differentiation for product formation can be induced and rigidly controlled. A solution to this problem may be found in developing a two-stage fermentor system as outlined by Noguchi et al.[15] In the first stage cell cultures grow in a turbidostat environment, while in the second stage they are switched to an environment that mimics the cytodifferentiating condition for product formation.

Chemostat cultures have been initiated with a cell line of *Acer pseudoplatanus* (Sycamore) under both nitrogen (nitrate) and phosphorus (phosphate) limitation. The potential of chemostat cultures for industrial purposes will certainly prompt further studies of limiting factors (macro-nutrients, phytohormones, light, temperature). In the production of metabolites by cell suspension cultures, chemostats may prove indispensable for stabilizing the cell population at the point of maximum product accumulation. Once the accumulation of secondary metabolites as a function of viable cells has been optimized, the harvest and processing of substances may be performed continually or intermittently.

Continuously synchronized plant cell suspension cultures can only be further improved when true single cell suspensions can be obtained, as the interrelationship with cells growing in aggregates will affect further response to synchronizing treatments. Single cells can readily be produced by transforming tissues or cells to protoplasts. Viable protoplasts, however, will resynthesize a cell-wall and upon division will reform aggregates. Improvements in culturing plant cells in continuous systems, therefore, is as much a technical, as it is a biological problem.

REFERENCES

1. Reinert, J. and Bajaj, Y. P. S., Eds., *Applied and Fundamental Aspects of Plant Cell, Tissue, and Organ Culture,* Springer Verlag, Berlin, 1977.
2. Murashige, T., Plant propagation through tissue culture, *Annu. Rev. Plant Physiol.,* 25, 135, 1974.
3. Furuya, T., Syono, K., Kojima, H., Hirotani, M., Ikuta, A., Hikichi, M., Kawaguchi, K., and Matsumoto, K., Chemical constituents and transformation capacity of medicinal plant callus tissues, in Proc. IV International Fermentation Symposium — Fermentation Technology Today, Kyoto, Terui, G., Ed., 1972, 705.
4. Alfermann, A. W., Bay, H. M., Doller, P. C., Hagedorn, W., Heins, M., Wahl, J., and Reinhard, E., Biotransformation of cardiac glycosides by plant cell cultures, in *Plant Tissue Culture and its Bio-technological Application,* Barz, W., Reinhard, E., and Zenk, M. H., Eds., Springer Verlag, Berlin, 1977, 125.
5. Street, H. E., Ed., *Plant Tissue and Cell Culture,* University of California Press, Berkeley, 1973.
6. Gamborg, O. L. and Wetter, L. R., *Plant Tissue Culture Methods,* National Research Council of Canada, Ottawa, 1975.
7. Murashige, T. and Skoog, F., A revised medium for rapid growth and bioassays with tobacco tissue cultures, *Physiol. Plant.,* 15, 473, 1962.
8. Gamborg, O. L., Miller, R. A., and Ojima, K., Nutrient requirements of suspension cultures of soybean root cells, *Exp. Cell Res.,* 50, 151, 1968.
9. Gamborg, O. L., Murashige, T., Thorpe, T. A., and Vasil, I. K., Plant tissue culture media, *In Vitro,* 12, 473, 1976.
10. Kao, K. N. and Michayluk, M. R., Nutritional requirements for growth of *Vicia hajastana* cells and protoplasts at very low population density in liquid media, *Planta,* 126, 105, 1975.

11. **King, P. J. and Street, H. E.**, Growth patterns in cell cultures, in *Plant Tissue and Cell Culture*, Street, H. E., Ed., University of California Press, Berkeley, 1973, 269.

12. **Fowler, M. W.**, Growth of cell cultures under chemostat conditions, in *Plant tissue culture and its Bio-technological Application*, Barz, W., Reinhard, E., and Zenk, M. H., Eds., Springer Verlag, Berlin, 1977, 253.

13. **Bayley, I. M., King, J., and Gamborg, O. L.**, The effect of the source of inorganic nitrogen on growth and enzymes of nitrogen assimilation in soybean and wheat cells in suspension cultures, *Planta*, 105, 15, 1972.

14. **Eriksson, T.**, Studies on the growth requirements and growth measurements of cell cultures of *Haplopappus gracilis, Physiol. Plant.*, 18, 976, 1965.

15. **Noguchi, M., Matsumoto T., Hirata, Y., Yamamoto, K., Katsuyama, A., Kato, A., Azechi, S., and Kato, K.**, Improvement of growth rates of plant cell cultures, in *Plant Tissue Culture and its Bio-technological Application*, Barz, W., Reinhard, E., and Zenk, M. H., Eds., Springer Verlag, Berlin, 1977, 85.

16. **Rose, D. and Martin, S. M.**, Parameters for growth measurements in suspension cultures of plant cells, *Can. J. Bot.*, 52, 903, 1974.

17. **Böhm, H.**, Secondary metabolism in cell cultures of higher plants and problems of differentiation, in *Molecular Biology, Biochemistry, and Biophysics*, Vol. 23, Kleinzeller, A., Springer, G.F., and Wittmann, H. G., Eds., Springer Verlag, Berlin, 1977, 104.

18. **Chu, Yaw-en, and Lark, K. G.**, Cell-cycle parameters of soybean (*Glycine max* L.) cells growing in suspension culture: suitability of the system for genetic studies, *Planta*, 132, 259, 1976.

19. **Davies, M.**, Polyphenol synthesis in cell suspension cultures of Paul's scarlet rose, *Planta*, 104, 50, 1972.

20. **Nimz, H., Ebel, J., and Grisebach, H.**, On the structure of lignin from soybean cell suspension cultures, *Z. Naturforsch.*, 30C, 442, 1975.

21. **Hahlbrock, K.**, Regulatory aspects of phenylpropanoid biosynthesis in cell cultures, in *Plant Tissue Culture and its Bio-technological Application*, Barz, W., Reinhard, E., and Zenk, M. H., Eds., Springer Verlag, Berlin, 1977, 95.

22. **Constabel, F., Shyluk, J. P., and Gamborg, O. L.**, The effect of hormones on anthocyanin accumulation in cell cultures of *Haplopappus gracilis, Planta*, 96, 306, 1971.

23. **Widholm, J. M.**, Anthranilate synthetase from 5-methyltryptophan susceptible and resistant cultured *Daucus carota* cells, *Biochim. Biophys. Acta*, 279, 48, 1972.

24. **Zenk, M. N., El-Shagi, H., Arens, H., Stöckigt, J., Weiler, E. W., and Deus, B.**, Formation of the indole-alkaloids serpentine and ajmalicine in cell suspension cultures of *Catharanthus roseus*, in *Plant Tissue Culture and its Bio-technological Application*, Barz, W., Reinhard, E., and Zenk, M. H., Eds., Springer Verlag, Berlin, 1977, 27.

25. **Kartnig, Th., Russheim, V., and Maunz, B.**, Beobachtungen uber das Vorkommen und die Bildung vou Cardenoliden in Gewebekulturen aus *Digitalis purpurea* und *Digitalis lanata, Planta Med.*, 29, 275, 1976.

26. **Reinhard, E.**, Biotransformations by plant tissue cultures, in *Tissue Culture and Plant Science*, Street, H. E., Ed., Academic Press, London, 1974, 433.

27. **Zenk, M. H., El-Shagi, H., and Schulte, V.**, Anthraquinone production by cell suspension cultures of *Morinda citrifolia, Planta Med.*, Suppl. 79, 1975.

28. **Steck, W., Gamborg, O. L., and Bailey, B. K.**, Increased yields of alkaloids through precursor biotransformation in cell suspension cultures of *Ruta graveolens, Lloydia*, 36, 93, 1973.

29. **D'Amato, F.**, Cytogenetics of differentiation in tissue and cell cultures, in *Plant Cell, Tissue, and Organ Culture*, Reinert, J. and Bajaj, Y. P. S., Eds., Springer Verlag, Berlin, 1977, 343.

30. **Singh, B. D., Harvey, B. L., Kao, K. N., and Miller, R. A.**, Selection pressure in cell populations of *Vicia hajastana* cultured in vitro, *Can. J. Genet. Cytol.*, 14, 65, 1972.

31. **Bayliss, M. W.**, The effects of growth in vitro on the chromosome complement of *Daucus carota* (L.) suspension cultures, *Chromosoma*, 51, 401, 1975.

32. **Withers, L. A. and Street, H. E.**, The freeze-preservation of plant cell cultures, in *Tissue Culture and its Bio-technological Application*, Barz, W., Reinhard, E., and Zenk, M. H., Eds., Springer Verlag, Berlin, 1977, 226.

33. **Kurz, W. G. W.**, A chemostat for single cell cultures of higher plants, in *Tissue Culture: Methods and Applications*, Kruse, P. F., Jr. and Patterson, M. K., Eds., Academic Press, New York, 1973, 359.

34. **Constabel, F., Kurz, W. G. W., Chatson, B., and Gamborg, O. L.**, Induction of partial synchrony in soybean cell cultures, *Exp. Cell Res.*, 85, 105, 1974.

35. **Constabel, F., Kurz, W. G. W., Chatson, K. B., and Kirkpatrick, J. W.**, Partial synchrony in soybean cell suspension cultures induced by ethylene, *Exp. Cell Res.*, 105, 263, 1977.

36. **Veliky, I. A. and Martin, S. M.,** A fermentor for plant cell suspension cultures, *Can. J. Microbiol.,* 16, 223, 1970.
37. **Miller, R. A., Shyluk, J. P., Gamborg, O. L., and Kirkpatrick, J. W.,** Phytostat for continuous culture and automatic sampling of plant-cell suspensions, *Science (N. Y.),* 159, 540, 1968.
38. **Wilson, S. B., King, P. J., and Street, H. E.,** Studies on the growth in culture of plant cells. XII. A versatile system for the large scale batch or continuous culture of plant cell suspensions, *J. Exp. Bot.,* 21, 177, 1971.
39. **Eriksson, T.,** Studies on the growth requirements and growth measurements of cell cultures of *Haplopappus gracilis, Physiol. Plant,* 18, 976, 1965.
40. **Eriksson, T.,** Partial synchronization of cell division in suspension cultures of *Haplopappus gracilis, Physiol. Plant.,* 19, 900, 1966.
41. **Eriksson, T.,** Duration of the mitotic cycle in cell cultures of *Haplopappus gracilis, Physiol. Plant.,* 20, 348, 1967.
42. **Eriksson, T.,** Effects of ultraviolet and X-ray radiation on in vitro cultivated cells of *Haplopappus gracilis, Physiol. Plant.,* 20, 507, 1967.
43. **Doree, M., Leguay, J-J., Terrine, C., Sadorage, P., Trapy, F., and Guern, J.,** Adaptation a l'adenine des cellules d'Acer pseudoplatanus: modalités d'utilisation de l'adénine exogene, in *Les Cultures de Tissus de Plantes,* Colloques Internationaux du Centre National de la Recherche Scientifique, 193, Paris, 345, 1971.
44. **Verma, D. and van Huystee, R.,** Derivation, characteristics, and large scale culture of a cell line from *Arachis hypogea* L. cotyledons, *Exp. Cell Res.,* 69, 402, 1971.
45. **Street, H. E., King, P. J., and Mansfield, K.,** Growth control in plant cell suspension cultures, in *Les Cultures de Tissus de Plantes,* Colloques Internationaux du Centre National de la Recherche Scientifique, 193, Paris, 17, 1971.
46. **Steck, W. and Constabel, F.,** Biotransformations in plant cell cultures, *Llyodia,* 37, 185, 1974.
47. **Kurz, W. G. W.,** A chemostat for growing higher plant cells in single cell suspension cultures, *Exp. Cell Res.,* 64, 476, 1971.
48. **Wilson, G.,** The Nutrition and Differentiation of Cells of *Acer pseudoplatanus* L. in Suspension Culture, Ph.D. thesis, University of Birmingham, England, 1971.
49. **King, P. J. and Street, H. E.,** Growth patterns in cell cultures, in *Plant Tissue and Cell Culture,* Street, H. E., Ed., University of California Press, Berkeley, 1973, 269.
50. **Fowler, M. W. and Clifton, A.** Activities of enzymes of carbohydrate metabolism in cells of *Acer pseudoplatanus* L. maintained in continuous (chemostat) culture, *Eur. J. Biochem.,* 45, 445, 1974.
51. **Young, M.,** Studies on the growth in culture of plant cells. Nitrogen assimilation during nitrogen-limited growth of *Acer pseudoplatanus* L. cells in chemostat culture, *J. Exp. Bot.,* 24, 1172, 1973.
52. **King, P. J.,** Studies on the growth in culture of plant cells. XX. Utilization of 2,4-dichlorophenoxyacetic acid (2,4-D) by steady-state cell cultures of *Acer pseudoplatanus* L, *J. Exp. Bot.,* 27, 1053, 1976.
53. **Wahl, J.,** Airlift-fermentater zur Züchtung von pflanzlichen Zellkulturzen, *GIT Fachz. Lab.,* 23, 169, 1979.

Chapter 7

USE OF CYCLOSTAT CULTURES TO STUDY PHYTOPLANKTON ECOLOGY

G-Yull Rhee, Ivan J. Gotham, and Sallie W. Chisholm

TABLE OF CONTENTS

I. INTRODUCTION

Phytoplankton have evolved in and adapted to an environment in which light and dark alternate periodically, and various factors regulate their growth. Phytoplankton are known to possess division cycles. These cycles may be entrained by a 24 hr photocycle in laboratory culture,[1-5] as they occur in the natural environment.[6-9] Thus physiological and ecological adaption should be considered a dynamic interrelation between environmental factors and the algal division cycle.

Experiments using steady-state chemostat cultures under continuous light have significantly advanced our knowlege of physiological changes under nutrient limitation. In this nonperiodic steady state, growth rate is independent of time and is determined only by dilution rate. Under these conditions, or if division cycles are disregarded, mathematical treatment of growth is simple. However, when a chemostat culture is illuminated for a part of a 24 hr period, growth is also a function of time during that photocycle. When growth of such a culture follows an identical pattern during each 24 hr cycle, it is termed a periodic cyclostat culture.[10] Studies of these cultures provide information on algal division cycles and thus deeper insights into the ecology of these organisms.

The ecological relevance of steady-state chemostat and cyclostat systems must be viewed within the context for which these systems are designed: nutrient-limited and time-independent growth. In cyclostats, growth is independent of time when averaged or integrated over the photocycle. However, growth rate in natural systems is neither time independent nor steady state. Thus the time-independent states of chemostat or cyclostat systems are, as stated by Jannasch,[11] utterly unnatural. A chemostat or a cyclostat culture is useful in phytoplankton ecology only to elucidate general physiological, biochemical, and genetic principles as related to ecology, rather than to reproduce the conditions existing in the natural environment.

Since the use of continuous culture to study phytoplankton ecology has been reviewed recently,[12] this article will concentrate on comparing the basic concepts of both chemostat and cyclostat systems.

II. GENERAL PRINCIPLES FOR CONTINUOUS CULTURE OF ALGAE

A. Chemostat

The specific rate of change in cell concentration (X) in a chemostat is the difference between specific growth rate (μ) and dilution rate (D). Thus,

$$d \ln X / dt = \mu - D \tag{1}$$

The specific growth rate is the rate of population increase per unit cell (or biomass) concentration $(1/X \cdot dX/dt)$. The reciprocal of dilution rate is the mean residence time of the population in the culture vessel. Both μ and D have units of time.$^{-1}$ For the reader's convenience a glossary of definitions for variables and constants is presented at the end of this chapter.

The rate of change in limiting nutrient concentration (S) in the vessel is the result of its rate of input $(D \cdot S_r)$ minus its rate of loss through out flow $(D \cdot S)$ and rate of uptake (V) by the population $(V \cdot X)$.

$$dS/dt = D(S_r - S) - V \cdot X \tag{2}$$

In a steady-state chemostat, μ and V are functions of the intra-cellular nutrient level, or the cell quota of limiting nutrient (q), and of S. In a cyclostat culture, however, they are also a function of time t during the period T, as discussed later in this section.

The rate of change in q can also be specified as the difference between V and rate of loss of q to daughter cells through division $(\mu \cdot q)$:

$$dq/dt = V - \mu \cdot q \tag{3}$$

Under constant environmental conditions in a chemostat, the population will ultimately achieve a self-regulating steady-state growth rate, and time-independent solutions for Equations 1 to 3 can be obtained:

$$\mu = D \tag{4}$$

and

$$V = \mu \cdot q = D(S_r - S)/X = D \cdot q \tag{5}$$

Equation 4 shows that steady-state growth of a chemostat culture is a strict function of dilution rate. Experimental designs of chemostat are based principally on this simple relationship. Thus the effects of nutrient limitation on cell physiology may be studied by changing the dilution rate. Alternatively, the effects of environmental conditions on nutrient-limited growth may be assessed by holding the dilution rate constant and varying systematically a single environmental parameter.

B. Cyclostat

Rates of various differentiation processes leading to cell division are neither time independent nor steady state.[13-16] Thus, in a cyclostat, the cell cycles of algae tend to be phased with the photocycle.

In cyclostat cultures only the period average of the time derivatives of S, X, V and μ are indpendent of time.[17,18]

$$\langle \mu \rangle_T = \langle V \cdot X \rangle_T / \langle q \cdot X \rangle_T = \langle V \cdot X \rangle_T / (S_r - \langle S \rangle_T) \tag{6}$$

$$\langle V \rangle_T = \langle \mu \cdot q \rangle_T \tag{7}$$

where $\langle \ldots \rangle_T = (1/T) \, _0\!\int^T \ldots \, dt$, or period average, and $T = 24$ hr.

From these equations

$$D \cdot T = \int_0^T \mu(t) \, dt \tag{8}$$

where $D \cdot T$ is integrated specific growth over period T and is numerically, but not dimensionally, equivalent to the dilution rate in units of day^{-1}. Therefore D determines only the integrated instantaneous growth rate $\mu(t)$ averaged over the period T, and, in contrast to the chemostat relationships in Equations 4 and 5, only the period average of nutrient uptake is equal to the period averages loss of q upon division. However, with these periodic steady-state kinetics can express population behavior ranging from nonperiodic steady–state, or random division, to completely synchronous division (Equations 7 to 8). Steady-state chemostat growth may thus be considered as a special, limiting case of the cyclostat system.[17,18]

To obtain increased biological and ecological information in cyclostat cultures, simple nonperiodic chemostat kinetics are replaced by somewhat more complex mathematics. Despite this complexity, there are empirical analogies between chemostat and cyclostat growth as will be seen in the following sections.

III. NUTRIENT-LIMITED GROWTH

A. Growth Rate and External Nutrient Concentration
1. Chemostat

For nutrient-limited growth of bacteria, Monod[19] first reported that μ and S are related by the empirically derived function

$$\mu = \bar{\mu}_m \cdot S/(K_s + S) \qquad (9)$$

where $\bar{\mu}_m$ is the theoretical maximum specific growth rate when $S \rightarrow \infty$ and K_s is the half-saturation constant, or the concentration of limiting nutrient required by the organism to achieve $\bar{\mu}_m/2$. The parameter $\bar{\mu}_m$ is greater than μ_m, the true physiological maximum growth rate, by a factor of $(1 + K_s/S_n)$, where S_n is the external concentration of a given nutrient when it is no longer limiting to growth. This equation resembles a steady-state equation for rates of enzyme catalyzed, homogeneous reactions (Michaelis-Menten enzyme kinetics). Thus K_s may be regarded as a species-specific parameter indicating the efficiency with which an organism utilizes a limiting substrate for growth. In algae this constant can be obtained directly by measuring μ and S, or may be approximated indirectly as a function of nutrient uptake, q, and $\hat{\mu}_m$ (Equations 26 and 27). When nutrients are at saturating levels, $\bar{\mu}_m$ will determine the outcome of competitive interactions. When nutrients are limiting, and the values of $\bar{\mu}_m$ for competing species is the same, however, K_s will determine the outcome. Values of K_s and $\bar{\mu}_m$ for a number of species and limiting nutrients are presented in Table 1.

In algal cultures it is difficult to assess the true value of K_s by direct measurement. The half-saturation constant for a limiting nutrient is generally lower than the analytical limit of detection. Furthermore, such factors as substrate-binding excretion products,[20,21] minimum substrate concentration for uptake,[23-25] and minimum dilute rates below which the principles of continuous culture do not apply,[22] further complicate experimental evaluation of this parameter.

2. Cyclostat

In the cyclostat environment, the phases of growth[10,17,18,26-29] and nutrient uptake[18,27,28,30] may be uncoupled during the period. Examples of light-entrained nutrient uptake and division in a number of laboratory and field populations of algae are shown in Table 2.

Since instantaneous growth rate $\mu(t)$ and instantaneous substrate level $S(t)$ vary periodically in a cyclostat, $\mu(t)$ and $S(t)$ cannot be related through the Monod function. However, in cultures of *Euglena gracilis* where phosphate is the limiting nutrient, the period average growth rate $\langle\mu\rangle_T$ can be related to the concentration of the residual limiting nutrient at the peak of its own oscillation (S_o) by a function similar to the Monod equation:[28]

$$\langle\mu\rangle_T = \langle\bar{\mu}_m\rangle_T \cdot S_0/(K_{so} + S_0) \qquad (10)$$

where S_o is the substrate concentration at the beginning of the daily or periodic division cycle of the population; $\langle\bar{\mu}_m\rangle_T$ is the theoretical maximum period averaged specific growth rate; and K_{so} is the half-saturation concentration in terms of S_o. This relation-

Table 1
HALF SATURATION CONSTANTS (K_s) AND THEORETICAL MAXIMUM SPECIFIC GROWTH RATES (μ_m) OF THE MONOD GROWTH FUNCTION FOR CONTINUOUS AND SEMICONTINUOUS CULTURES OF ALGAE

Species	Limiting nutrient	K_s (μmoles$\cdot$L^{-1})	μ_m (day^{-1})	Light cycle	Ref.
Asterionella for-	P	0.02	0.88	14/10	1
mosa	Si	3.94	1.06	14/10	1
Chlorella pyrenoi- dosa	NO₃	0.43	0.98[a]	Continuous	2
Cyclotella meneghi-	P	0.25	0.78	14/10	1
niana	Si	1.44	1.33	14/10	1
Diatoma elongatum	P	0.02	0.70	14/10	3
	Si	1.51	1.20	14/10	3
Euglena gracilis	P	0.02	0.69[b]	14/10	4
Monochrysis lutheri	P	0.04[a]	0.97[a]	Continuous	5
Nitzschia actinaster- oides	P	0.01	2.09	Continuous	6
Oscillatoria agardhii	P	0.03	0.5	Continuous	7
	NO₃	0.04	0.5	Continuous	7
Scenedesmus sp.	P	0.011	1.3	12/12	8
	NO₃	1.0	1.3	Continuous	9
Thalassoisira pseu- donana	B₁₂	1.92 × 10⁻⁷	1.20	Continuous	10

Note: Summary of half-saturation constants (K_s) and theoretical maximum specific growth rates ($\bar{\mu}_m$) for representative species and limiting nutrient for Monod growth function. See glossary and text for further definitions of species parameters.

[a] Average of author's data.
[b] Integrated growth.

REFERENCES

1. **Tilman, D. and Kilham, S. S.**, Phosphate and silicate growth and uptake kinetics for the diatoms *Asterionella formosa* and *Cyclotella meneghiniana* in batch and semi-continuous culture, *J. Phycol.*, 12, 375, 1976.
2. **Pickett, J. M.**, Growth of *Chlorella* in a nitrate limited chemostat, *Plant Physiol.*, 55, 223, 1975.
3. **Kilham, S. S., Kott, C. L., and Tilman, D.**, Phosphate and silicate kinetics for the Lake Michigan diatom *Diatoma elongatum*, *J. Great Lakes Res.*, 3, 93, 1977.
4. **Gotham, I. J.**, Nutrient Limited Cyclostat Growth: a Theoretical and Physiological Analysis, Ph.D. thesis, University of New York, Albany, 1977.
5. **Goldman, J. C.**, Steady-state growth of phytoplankton in continuous culture: comparison of internal and external nutrient models, *J. Phycol.*, 13, 251, 1977.
6. **Muller, H.**, Das Wachstum von *Nitzschia actinasteroides* (Lemm.)V. Goor im chemostaten bei limitierender phosphatkonzentration, *Ber. Dtsch. Bot. Ges.*, 83, 537, 1970.
7. **Ahlgren, G.**, Growth of *Oscillatoria agardhii* Gom. in chemostat culture. I. Investigation of nitrogen and phosphorus requirements, *Oikos*, in press, 1978 .
8. **Rhee, G-Y.**, A continuous culture study of phosphate uptake, growth rate, and polyphosphate in *Scenedesmus* sp., *J. Phycol.*, 9, 495, 1973.
9. **Rhee, G-Y.**, Phosphate uptake under nitrate limitation by *Scenedesmus* sp. and its ecological implications, *J. Phycol.*, 10, 470, 1974.
10. **Swift, D. G. and Taylor, W. R.**, Growth of vitamin B₁₂-limited cultures: *Thalassiosira pseudonana*, *Monochrysis lutheri*, and *Isochrysis galbana*, *J. Phycol.*, 10, 385, 1974.

Table 2

Species	Oscillating Variable	Phase Relationships	Ref.
Asterionella formosa	Division Cell-P quota Cell-N quota P-Uptake		1
Chaetoceros gracilis	Division Si-Uptake Si-Incorporation		2
Ditylum brightwellii	Division Si-Incorporation Si-Uptake		2
Ditylum brightwellii	Division Cell-N quota Cell-P quota		3
Euglena gracilis	Division P-Uptake Cell-P quota		4—7
Monochrysis lutheri	Division N_a-Uptake N_o-Uptake		8
Scenedesmus obtusiusculus	Division P-Uptake		9
Scenedesmus sp.	Division P-Uptake		10
Skeletonema costatum	Division Si-Uptake Si-Incorporation		2

Table 2 (continued)

Species	Oscillating Variable	Phase Relationships	Ref.
Skeletonema costatum	Division Cell-N quota N_a-Uptake N_o-Uptake	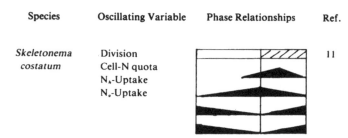	11

Note: Examples of phase relationships for nutrient uptake and incorporation, cell division and cell quota for algae grown on 24 hr light/dark cycles. Hatched areas represent dark period. The base of the solid triangles depict approximate period over which the oscillating variable occurs. Apex of triangles indicate approximate time in maxima for rate or quota. N_o and N_a are nitrate and ammonium, respectively.

REFERENCES

1. Gotham, I. J. and Rhee, G-Y., manuscript in preparation, 1980.
2. Chisholm, S. W., Azam, F., and Eppley, R. W., Silicic acid incorporation in marine diatoms on light dark cycles: use as an assay for phased cell division, *Limnol. Oceanogr.*, 23, 518, 1978.
3. Eppley, R. W., Holmes, R. W., and Paasche, E., Periodicity in cell division and physiological behavior of *Ditylum brightwellii*, a marine planktonic diatom, during growth in light-dark cycles, *Arch. Mikrobiol.*, 56, 305, 1967.
4. Chisholm, S. W., Stross, R. G., and Nobbs, P. A., Light/dark phased cell division in *Euglena gracilis* (Z) (Euglenophyceae) in PO_4-limited continuous culture, *J. Phycol.*, 11, 367, 1975.
5. Chisholm, S. W. and Stross, R. G., Phosphate uptake kinetics in *Euglena gracilis* (Z) (euglenophyceae) grown in light/dark cycles. I. Synchronized batch cultures, *J. Phycol.*, 12, 210, 1976.
6. Chisholm, S. W. and Stross, R. G., Phosphate uptake kinetics in *Euglena gracilis* (Z) (Euglenophyceae) grown in light/dark cycles II. Phased PO_4-limited cultures, *J. Phycol.*, 12, 217, 1976.
7. Gotham, I. J., Nutrient Limited Cyclostat Growth: a Theoretical and Physiological Analysis, Ph.D. thesis, University of New York, Albany, 1977.
8. Caperon, J. and Ziemann, D. A., Synergistic effects of nitrate and ammonium ions on the growth and uptake kinetics of *Monochrysis lutheri* in continuous culture, *Mar. Biol.*, 36, 73, 1976.
9. Sundberg, I. and Nilshammar-Holmvall, M., The diurnal variations and phosphate uptake and ATP level in relation to deposition of starch, lipid, and polyphosphate in synchronized cells of *Scenedesmus*, *Z. Pflanzenphysiol.*, 76, 270, 1975.
10. Azad, H. S. and Borchardt, J. A., Variations in phosphorus uptake by algae, *Environ. Sci. Technol.*, 4, 737, 1970.
11. Eppley, R. W., Rogers, J. N., McCarthy, J. J., and Sournia, A., Light/dark periodicity in nitrogen assimilation of the marine phytoplankters *Skeletonema costatum* and *Coccolithus huxleyi* in N-limited chemostat cultures, *J. Phycol.*, 7, 150, 1971.

ship indicates that once cell division is initiated by a cyclostat population of *Euglena gracilis*, the process is completed according to a predetermined sequence that is little influenced by limiting substrate levels. The amplitude of the oscillation of X, which is related to $\langle\mu\rangle_T$, increases with increasing D.[10,28] In a numerical simulation of cyclostat growth for *Euglena gracilis*, Chisholm and Nobbs[31] have postulated that integrated growth $(D \cdot T)$ for the period is a function of the numerical average of the peak and trough of the oscillation in $S(\bar{S})$,

$$D \cdot T = \ln\left[\frac{(e^{D_{max} \cdot T} - 1) \cdot \bar{S}}{\bar{K}_s + \bar{S}} + 1\right] \tag{11}$$

where $D_{max} \cdot T$ is the maximum integrated growth and $\bar{K}_s$ is the half-saturation concentration of limiting substrate when $(e^{DT} - 1) = (e^{D_{max} \cdot T} - 1)/2$.

B. Growth and Cell Quota

1. Chemostat

Growth rates of phytoplankton are also directly related to cell quota (q) or to the concentration of specific storage pools. This relationship has been expressed by three empirical equations derived from chemostat studies. For *Isochrysis galbana*:[32]

$$\mu = \hat{\mu}_m \cdot (q - q_0)/(K_q + (q - q_0)) \tag{12}$$

For *Monochrysis lutheri*:[20]

$$\mu = \hat{\mu}_m(1 - q_0/q) \tag{13}$$

For *Thalassiosira pseudonana* and *Thalassiosira fluviatilis*:[33]

$$\mu = \hat{\mu}_m(1 - 2^{-(q - q_0)/q}) \tag{14}$$

In these equations $\hat{\mu}_m$ is the theoretical maximum growth rate, K_q is the half saturation constant for growth as a function of q, and q_0 is the subsistence quota, or the value of q when $\mu = 0$.

The parameter K_q in Equation 12 is not significantly different from q_0, irrespective of organisms or limiting nutrient.[12,47] Thus Equation 12 is identical with Equation 13. Droop[21] found that the data of Fuhs[33] used to derive Equation 14 would also adequately fit Equation 13. Droop's equation (Equation 13) is, therefore, the simplest and apparently most universally applicable expression of growth in terms of cell quota. This function is linear if $D \cdot q$ is plotted against q; the slope is $\hat{\mu}_m$, and the intercept on the q axis is q_0. This model also appears suitable for batch cultures,[34] semicontinuous cultures,[35] and certain natural populations.[36,37] The applicability of the Droop model to transient states may be determined largely by the degree of perturbation to the culture and by the length of the lag time required for the cells to adapt metabolically to environmental changes.[12] It has been suggested, therefore, that the growth model applies to transient states only when cells require little time to readjust to the perturbation and environmental changes.[12] Representative examples of q_0 and μ_m for a number of species and limiting nutrients are presented in Table 3.

It was later recognized that the theoretical maximum specific growth rate $\hat{\mu}_m$ is larger than μ_m, the true physiological maximum growth rate, because q can never be infinite.[21] Thus μ_m is less than $\hat{\mu}_m$ by a factor of $(1 - q_0/q_m)$,[40-42] in which q_m is the cell quota when q is independent of external substrate levels or independent of μ.[12]

Table 3
SUBSISTANCE QUOTAS(q_o) AND THEORETICAL MAXIMUM GROWTH
RATE ($\hat{\mu}_m$) FOR DROOP'S GROWTH FUNCTION

Species	Limiting nutrient	q_o ($10^{-9}\mu$mole $\cdot$ cell^{-1})	μ_m (day^{-1})	Light cycle	Ref.
Anabaena flos-aquae	P	2.63	0.89	12/12	1
Asterionella for-mosa	P	1.75	0.70	14/10	2
	Si	296.	1.21	14/10	2
Chaetoceros gracilis	NH$_4$	26.0	3.07	Continuous	3
Cyclotella meneghi-niana	P	10.7	0.69	14/10	2
	Si	157.	1.16	14/10	2
Diatoma elongatum	P	3.24	0.72	14/10	4
	Si	384.	0.93	14/10	4
Dunaliella tertiolecta	NO$_3$—NH$_4$	60.	1.8	Continuous	5
Euglena gracilis	P	89.	0.69[b]	14/10	6,7
Isochrysis galvana	NO$_3$	30.	0.91	Continuous	8,9
Microcystis sp.	P	1.58	1.15	12/12	1
Monochrysis lutheri	P	0.37[a]	1.23[a]	Continuous	10
	B$_{12}$	2.48 $\cdot$ 10^{-3}[a]	1.14[a]	Continuous	10
Nitzschia actinaster-oides	P	2.74	2.09	Continuous	11
Scenedesmus sp.	P	1.6	1.35	12/12	12
	N	45.4	1.3	Continuous	13
Skeletonema costa-tum	B$_{12}$	8.81 $\cdot$ 10^{-4}	1.27	Continuous	9
Thalassiosira fluvia-tilis	P	12.5	1.11	Continuous	14
Thalassiosira pseu-donana	P	0.90	1.11	Continuous	14
	Si	20.	3.43	Continuous	15

Note: Summary of theoretical maximum specific growth rates ($\hat{\mu}_m$) and subsistence quotas(q_o) of the Droop growth model for representative species and limiting nutrients. See glossary and text for further definition of species parameters.

[a] Average of author's data
[b] Integrated growth

REFERENCES

1. Gotham, I. J. and Rhee, G-Y., Comparative kinetic studies of phosphate limited growth and phosphate uptake in phytoplankton in continuous culture, manuscript in preparation.
2. Tilman, D. and Kilham, S. S., Phosphate and silicate growth and uptake kinetics of the diatoms *Asterionella formosa* and *Cyclotella meneghiniana* in batch and semi-continuous culture, *J. Phycol.,* 12, 375, 1976.
3. Thomas, W. H. and Dodson, A. N., On nitrogen deficiency in tropical Pacific Oceanic phytoplankton. II. Photosynthetic and cellular characteristics of a chemostat-grown diatom, *Limnol. Oceanogr.,* 17, 515, 1972.
4. Kilham, S. S., Kott, C. L., and Tilman, D., Phosphate and silicate kinetics for the Lake Michigan diatom *Diatoma elongatum, J. Great Lakes Res.,* 3, 93, 1977.
5. Bienfang, P. K., Steady-state analysis of nitrate-ammonium assimilation by phytoplankton, *Limnol. Oceanogr.,* 20, 402, 1975.
6. Chisholm, S. W., Stross, R. G., and Nobbs, P. A., Light/dark phased cell division in *Euglena gracilis*(Z) (Euglenophyceae), in PO$_4$-limited continuous culture, *J. Phycol.,* 11, 367, 1975.
7. Gotham, I. J., Nutrient limited cyclostat growth. I. Empirical growth functions for *Euglena gracilis* (Z) (Euglenophyceae), manuscript in preparation, 1978.

8. **Caperon, J.**, Population growth response of *Isochrysis galbana* to nitrate variation at limiting concentrations, *Ecology*, 49, 867, 1968.
9. **Droop, M. R.**, Vitamin B_{12} and marine ecology. V. Continuous culture as an approach to nutritional kinetics, *Helgol. Wiss. Meeresunters.*, 20, 629, 1970.
10. **Droop, M. R.**, The nutrient states of algal cells in continuous culture, *J. Mar. Biol. Assoc. U.K.*, 54, 825, 1974.
11. **Muller, H.**, Wachstum und phosphate bedarf von *Nitzschia actinasteriodes* (Lemm.)V. Goor in statischer und homokontinuierlicker Kulter unter phosphatlimitierung, *Arch.Hydrobiol.*, *Suppl.*, 38, 399, 1972.
12. **Rhee, G-Y.**, A continuous culture study of phosphate uptake, growth rate, and polyphosphate in *Scenedesmus* sp., *J. Phycol.*, 9, 495, 1973.
13. **Rhee, G-Y.**, Phosphate uptake under nitrate limitation by *Scenedesmus* sp. and its ecological implications, *J. Phycol.*, 10, 470, 1974.
14. **Fuhs, G. W.**, Phosphorus content and rate of growth in the diatoms *Cyclotella nana* and *Thalassiosira fluviatilis*, *J. Phycol.*, 5, 312, 1969.
15. **Paasche, E.**, Silicon and the ecology of marine plankton diatoms.1. *Thalassiosira pseudonana* Hasle and Heimdal (*Cyclotella nana* Hustedt) grown in a chemostat with silicate as limiting nutrient, *Mar. Biol.*, 19, 117, 1973.

2. Cyclostat

In cyclostat cultures cell number, cell quota, and division rate oscillate over a 24 hr period (Figure 1 and Table 2). Oscillations in cell quota have been demonstrated in *Ditylum brightwellii*,[43] *Euglena gracilis*[10,18] and *Selenastrum*.[29] Chisholm et al.[10] have shown that the integrated growth in phosphate-limited cyclostat cultures of *Euglena gracilis* follows the Droop growth model if q is replaced by an average cell quota for the period. Its value is calculated by assuming that residual phosphate in the culture is negligible and dividing the concentration of inflowing phosphate by the average cell density for the oscillation, or $(X_{max} + X_{min})/2$. Eppley and Renger[44] have also employed arithmetic averages to describe nitrogen limited, periodic growth of *Thalassiosira pseudonana* in a chemostat, but these averages obscure the ecologically important rhythmic properties of nutrient-limited populations.

Since $\mu(t)$ and instantaneous q values, $q(t)$, vary periodically in cyclostat cultures, plots of $\mu(t)$ against $q(t)$ for any given growth cycle describe closed trajectories[17,18] (Figure 1). At the beginning of the growth cycle of *Euglena gracilis* (1 hr after lights go on), $q(t)$ is at the trough of its oscillation and $\mu(t)$ is zero (Figure 1, point a_1) because cell division is restricted to a specific time interval during the light/dark cycle. Since $\mu(t)$ is zero during the remainder of the light phase, $q(t)$ increases through nutrient uptake (Equation 3) from the minimum (q_c) to a maximum for the period (Figure 1, point a_2). Immediately after the beginning of the dark phase of the photocycle (14 hr after lights go on), a division burst occurs (Figure 1 point a_3). Afterward $\mu(t)$ declines by an apparent pseudo-first-order process throughout the dark phase and for 1 hr of the next light phase.[18] As a result, $q(t)$ decreases (Equation 3) throughout the division phase by net loss to daughter cells (Figure 1, points a_3 to a_4). At the end of the division phase (1 hr after lights go on) $\mu(t)$ becomes zero and $q(t)$ returns to the initial condition (q_c).

Apparently q_c can be related to period average growth rate. This relationship is described by a function similar to Equation 12:[18,28]

$$\langle \mu \rangle_T = \frac{\langle \hat{\mu}_m \rangle_T \cdot (q_c - q_{co})}{K_{qc} + (q_c - q_{co})} \tag{15}$$

where $q_{co} = q_c$ when $\mu = 0$, K_{qc} is the half-saturation constant, and $\langle \hat{\mu}_m \rangle_T$ is the theoretical maximum specific period average growth rate.

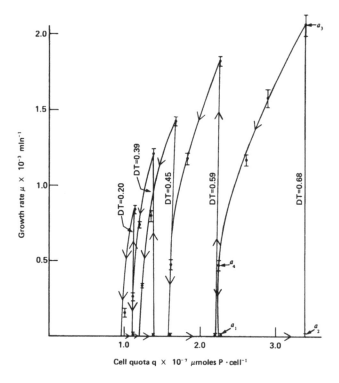

FIGURE 1. Phase space plots of instantaneous growth rate ($\mu(t)$) against cell quota ($q(t)$) for several integrated growths (DT) of P-limited *Euglena gracilis*. Solid lines are the output from a simulation model.[104] Dots with 95% confidence limits are experimental data.[105] See text for further details.

It is clear from Figure 1 that the efficiency of nutrient uptake, or the buildup of q_c during the nondividing phase, may be an important factor in determining the result of interaction among species or in allowing their coexistence. It is also evident that this factor cannot be elucidated in chemostat cultures and that instantaneous growth rate in a cyclostat cannot be described by arithmetic averages of instantaneous values of $q(t)$ over the light/dark cycle.

If continuous illumination replaced the light/dark cycle, cell division and nutrient uptake of the cells would ultimately become random and each loop in Figure 1 would spiral inward to a single point. These nodal points would follow the steady-state equation of Droop (Equation 13).[10]

IV. NUTRIENT UPTAKE, FEEDBACK AND DERIVATION OF K_s

Since nutrient uptake in steady-state chemostat cells has been discussed in detail elsewhere,[12] in addition to uptake under periodic conditions, nonperiodic uptake will be considered here only as it applies to the derivation of K_s in Droop's growth model.

In his original derivation, Droop[20] treated nutrient-uptake rate as independent of cell quota or growth rate. Rhee,[45] however, found that in phosphate-limited *Scenedesmus* sp. the maximum uptake rate of this nutrient increases as μ (thus q) decreases. Thus

$$V = \hat{V}_m \cdot S/(K_m + S) \qquad (16)$$

where $\hat{V}_m$ is the apparent maximum uptake velocity at a given μ or q and K_m is the half-saturation constant for uptake. This feedback effect was subsequently found in *Euglena gracilis*,[28,30] *Thalassiosira pseudonana*,[46] *Anabaena flos-aquae*,[47] *Microcystis* sp.,[47] and *Asterionella formosa*.[47] Feedback effects of q on nitrate uptake have also been observed in nitrate-limited *Scenedesmus* sp.,[39] *Asterionella formosa*,[47] and *Thalassiosira pseudonana*.[44]

In experiments with *Scenedesmus* sp.,[45] *Asterionella formosa*,[47] *Microcystis* sp.,[47] and *Anabaena flos-aquae*,[47] K_m was not affected by q or μ. Thus phosphate uptake formally resembles kinetics of noncompetitive enzyme inhibition. Inhibition can be expressed by intracellular polyphosphate levels or by surplus-P (hot-water-extractable P).[38,45]

$$V = \frac{V_m}{(1 + K_m/S)(1 + i/K_i)} \tag{17}$$

where V_m is the true maximum rate of uptake, which is obtained when $\mu = 0$, i is the cellular level of inhibitor and K_i is the inhibition constant. The relationship between $\hat{V}_m$ and i is

$$1/\hat{V}_m = 1/V_m + i/V_m \cdot K_i \tag{18}$$

Thus, if $1/\hat{V}_m$ in Equation 18 is plotted against i, the intercepts of the ordinate and abscissa are the values of $1/V_m$ and $-K_i$. Equation 18 implies that the homeostatic balance maintained between i and V_m keeps V_m constant. The relationship of these terms may be expressed as

$$1/\hat{V}_m \cdot Z = 1/V_m \tag{19}$$

where $Z = 1 + i/K$.

When V_m in Equation 17 is expressed in terms of Z in Equation 19

$$1/V = (K_m/\hat{V}_m \cdot Z)(1/S) + 1/\hat{V}_m \cdot Z \tag{20}$$

From Equation 13

$$1 = \mu/\hat{\mu}_m + q_0/q, \tag{21}$$

and since $q = V/\mu$ (Equation 5),

$$1/\mu = 1/\hat{\mu}_m + q_0/V. \tag{22}$$

When Equations 20 and 22 are combined,

$$1/\mu = 1/\hat{\mu}_m + q_0 \left(\frac{K_m}{\cdot \hat{V}_m \cdot Z} \cdot \frac{1}{S} + \frac{1}{\hat{V}_m \cdot Z} \right) \tag{23}$$

After rearranging and dividing by $\hat{V}_m \cdot Z$,

$$\mu = \frac{\hat{\mu}_m \cdot S}{\hat{\mu}_m \cdot q_0 \cdot K_m/\hat{V}_m \cdot Z + S(\hat{\mu}_m \cdot q_0/\hat{V}_m \cdot Z + 1)} \tag{24}$$

Since $\hat{V}_m \cdot Z$ is much greater than $\hat{\mu}_m \cdot q_o$,[38,47] Equation 24 reduces to

$$\mu = \frac{\hat{\mu}_m \cdot S}{\hat{\mu}_m \cdot q_o \cdot K_m / \hat{V}_m \cdot Z + S} \tag{25}$$

Provided that $\bar{\mu}_m / \bar{\mu}_m \approx 1$, then this is analogous to the Monod growth function where

$$K_s = \hat{\mu}_m \cdot q_o \cdot K_m / \hat{V}_m \cdot Z \tag{26}$$

or

$$K_s = \hat{\mu}_m \cdot q_o \cdot K_m / V_m \tag{27}$$

Equation 26 shows that if the inhibition term Z is neglected, the value of K_s would vary depending on the value of i (and hence growth rate) at which $\hat{V}_m$ is determined. K_s is, however, a species-specific constant that must not vary. Thus when $\hat{V}_m$ values at growth rates other than $\mu = 0$ are used, corresponding values of the inhibition term must be known. Since there is no feedback in uptake in completely starved cells (or when $\mu = 0$), $\hat{V}_m = V_m$ and no inhibition term is required (Equation 27).

When K_s is determined directly in a dilute batch culture,[48] the inoculum must also be completely starved. If the cells in the inoculum had intracellular nutrient at levels above subsistence quota, growth would occur even when $S = 0$ and this would result in an overestimation of K_s.[12]

Since K_s must remain constant while $\hat{V}_m$ and Z vary, Equation 26 implies that as the degree of starvation increases, so does the rate of uptake. This relationship, which is also apparent in Equation 17, seems to indicate an evolutionary development of a compensatory mechanism between nutrient uptake and intracellular nutrient concentrations. This mechanism may ensure survival in nutrient-limited environments.

At low concentrations of limiting substrate, K_s must be low to maintain highly efficient growth. In various organisms this seems to be accomplished by positive correlations between V_m and q_o under phosphate limitation (Equations 26 and 27 and Table 4). For example, if q_o is large, K_s increases; but if V_m is high, K_s decreases. The positive relationships indicate that the disadvantages of high subsistence requirements (q_o) may be offset by high maximum uptake capacity (V_m) for limiting nutrients.

At low growth rates, $\hat{V}_m$ increases as Z decreases. Therefore, if nutrients are extremely limiting to growth, a high V_m and hence high growth potential can be maintained. If S increases slightly for a short time for such cells, the limiting substrate may be taken up at a higher rate. Thus, as growth rate decreases, relaxation of feedback inhibition may increase the growth potential, i.e., the potential for a large increase in relative growth rate (μ/μ_m) at a given increase in substrate concentration.

In phosphate-limited cells, $\hat{V}_m$ for phosphate uptake is inversely related to the level of intracellular polyphosphate pools. In nitrogen-limited cells, $\hat{V}_m$ for nitrogen uptake has an inverse relationship with intracellular pools of free amino acids. But it is not clear whether the half-saturation constant for nitrogen uptake K_m remains unchanged as it does for phosphate uptake.[39] However, unless the $K_m / \hat{V}_m$ ratio remains constant and until nitrogen-uptake kinetics are clearly elucidated, K_i for nitrogen-limited growth should be determined the same way as for phosphate-limited growth. This would provide a basis for its comparison between various species.

In cyclostat cultures of *Euglena gracilis*, Gotham[28] found that Equations 26 and 27 also apply to periodic growth if $\hat{\mu}_m$ is expressed as $\langle\hat{\mu}_m\rangle_T$ and q_o as q_{oc} and if K_m, K_i

Table 4
KINETIC CONSTANTS FOR P UPTAKE AND P-LIMITED GROWTH

Species	$\hat{\mu}_m$ (day⁻¹)	q_o (×10⁻⁹ μmole·cell⁻¹)	V_m^a (×10⁻¹⁰ μmole·cell⁻¹·min⁻¹)	K_m (μM)	K_i^a (10⁻⁹ μmole·cell⁻¹)	K_s^b (μM)
Anabaena flos-aquae[1]	0.89	2.63	3.54	1.15	1.14	0.010
Asterionella formosa[1]	1.01	4.56	7.15	0.60	1.25	0.003
Microcystis sp.[1]	1.15	1.58	1.06	1.61	1.17	0.019
Scenedesmus sp.[2]	1.35	1.60	0.80	0.60	0.70	0.011

Note: Summary of kinetic parameters for P-limited growth in 5 species of various taxonomic levels: where $\hat{\mu}_m$ is the theoretical maximum specific growth rate; q_o the cell phosphate quota when $\mu = 0$; V_m the actual maximum uptake velocity when $i = 0$; K_m the half-saturation constant for uptake as a function of substrate K_i is the inhibition constant based on surplus-P (hot-water-extractable P); and K_s is the calculated half-saturation constant for growth as a function of extracellular substrate concentration. See glossary and text for further definition of species parameters.

a K_i and V_m based on cell surplus-P

b $K_s = q_o \cdot \hat{\mu}_m \cdot K_m / V_m$

REFERENCES

1. **Gotham, I. J. and Rhee, G-Y.**, Comparative kinetics studies of phosphate limited growth and phosphate uptake in phytoplankton in continuous culture, *J. Phycol.*, in press, 1981.
2. **Rhee, G-Y.**, A continuous culture study of phosphate uptake, growth rate, and polyphosphate in *Scenedesmus* sp., *J. Phycol.*, 9, 485, 1974.

and $\hat{V}_m$ or V_m are determined at the peak of the cycles for substrate oscillations. K_t in this case (K_{to} in Equation 10) is expressed as substrate concentrations at the peak of substrate oscillation.

Nutrient uptake oscillates with the same periodicity as the 24 hr light/dark cycle (Table 2). In phosphate-limited cultures of *Euglena gracilis*[28,30,50] and in natural assemblages of *Fragilaria* sp.[51] and *Tabellaria* sp.,[51] V_m for phosphate uptake oscillates, but not necessarily in phase with division (Table 2). The oscillation of V_m is not the result of feedback by intracellular pools of phosphate,[28] but seems to be due to a circadian biochemical oscillator within the cell. Edmunds[52,53] has shown that relative activities of many enzymes oscillate with a 24 hr frequency in nondividing populations. However, when the kinetic terms are measured in a cyclostat population at the trough of its oscillation and substituted in Equation 17, phosphate uptake can be described.[28] It is interesting that this expression is analogous to phosphate-limited growth in a cyclostat (Equations 10 and 15) in which $\langle\mu\rangle_T$ is determined by the cell quota at the trough of its oscillation or phosphate level at the peak of its oscillation.

Since $\mu(t)$, $V(t)$, and $q(t)$ can be out of phase in a 24 hr light/dark cycle, only the period averages of $V(\langle V\rangle_T)$ and $\mu \cdot q$ ($\langle\mu \cdot q\rangle_T$) are equal.[17,18,28] The temporal uncoupling occurs when DNA replication and cell division are out of phase with such physiological processes as photosynthesis, macromolecular synthesis, and respiration.[28,53-55]

V. CELLULAR NUTRIENT POOLS AND BIOSYNTHESIS

A. Chemostat

Algae accumulate many extracellular nutrients — among them phosphate, nitrogen, vitamin B_{12}, and silicon — in excess of immediate requirement for growth.[12] Under nonperiodic chemostat conditions, the relationship between the size of these pools and growth rate is similar to Droop's cell quota growth model (Equation 13). The physiological explanation of this model may lie in the relationship between pool sizes and growth rate, since cell quotas are proportional to pool sizes in the steady state.[38,39,45] These pools, however, appear to affect growth rate only indirectly through nucleic acid synthesis. There is a strong correlation between total RNA and polyphosphate concentrations in phosphate-limited cultures where nucleic acid synthesis is limited by the availability of phosphate from the intracellular pools,[45,56-60] and between RNA and free amino acid concentrations in nitrogen-limited cultures.[12,39]

As mentioned above, the size of intracellular pools may also regulate nutrient uptake by a feedback mechanism. Since the inhibition of phosphate uptake by polyphosphates is noncompetitive type, the feedback probably affects the amount of enzyme available for phosphate uptake rather than its affinity for substrate. A probable mechanism for this the noncompetitive inhibition is suggested by electron micrographs of *Diatoma* sp.[61] in which polyphosphates appear to condense around polysome-like particles, which gradually become electron-opaque toward the center as condensation increases.[103] Hence, as growth rate (or polyphosphate concentration) increases, synthesis of enzymes (possibly those responsible for phosphate uptake) should decrease.

B. Cyclostat

During a light/dark cycle in a cyclostat, on the other hand, oscillations in intracellular pools of polyphosphates have been demonstrated in *Euglena gracilis*,[28] *Ankistrodesmus braunii*,[62] *Chlorella elipsoidea*,[57] *Chlorella pyrenoidosa*,[56] and *Scenedesmus obstuisiusculus*.[63]

In *Scenedesmus obstuisiusculus* and *Euglena gracilis* maximum phosphate uptake is in phase with maximum polyphosphate pools. Thus under periodic illumination the rhythm in phosphate uptake is apparently unaffected by the level of polyphosphates. This may be explained by circadian rhythmicity in phosphate uptake, i.e., the persistence of rhythmic uptake demonstrated in phosphate limited, stationary-phase cultures of *Euglena gracilis* maintained on light/dark cycles.[30] Since polyphosphates approach zero as $<\mu>_T$ approaches zero,[28] it is likely that in this study[30] cellular polyphosphates were zero and unlikely that the rhythmic uptake was a function of feedback from oscillating pools of limiting nutrient.

In cyclostat cultures of *Euglena gracilis* the absolute value of $\hat{V}_m$ decreases as the growth rate[30,31,50] and polyphosphate concentrations increase at the trough of its daily growth oscillation.[28] However, the relative amplitude for the oscillation of $\hat{V}_m$ increases as the period-averaged growth rate increases. In addition, the peak and trough of the oscillation in $\hat{V}_m$ maintain a constant phase relative to the time of onset of cell division, but are independent of the period-averaged growth rate.[28] In this organism, the minima for the daily oscillation in polyphosphates apparently determine only the magnitude of $\hat{V}_m$ and not the rhythm in $\hat{V}_m$ itself.

Temporal uncoupling of nutrient uptake and division by internal pools could promote the coexistence of species that have uptake maxima for a common limiting nutrient at different times of the subjective day. As uptake reaches its peak, nutrient could be stored in intracellular pools, from which it could be withdrawn for biosynthesis during the remainder of the cycle. Thus uptake of the limiting nutrient by each species during a limited period of the photocycle would provide it with enough substrate for metabolic processes that are out of phase with uptake. Indeed, as discussed below, this is the case in two photosynthetic sulfur bacteria living in the same habitat. Mathematical models for optimal strategy for metabolism of intracellular reserves in microorganisms have been presented by Cohen and Parnas.[64,65]

During the light period the rate of carbon assimilation per cell also undergoes diel oscillations. The rate may change as much as sixfold from peak to trough in nitrate-limited cultures of *Coccolithus huxleyi*,[27] *Skeletonema costatum*,[27] and *Dunaliella tertiolecta*[66] and *Ditylum brightwellii*.[43] This oscillation is not due to changes in photosynthetic unit size, assimilation number, nor to the amount of cellular chlorophyll *a*.[67,68] Rather it seems to be related to changes in quantum efficiency or to the number of active photosynthetic units.[68] The timing of the activation of photosynthetic units relative to the light/dark cycle and to the cell's daily growth cycle implies an ordered sequence of metabolic functions that are under the control of a master cellular "clock", the underlying mechanisms for which are unclear. Endogenous circadian regulation of photosynthetic capacity has been observed in *Gonyaulax polyedra*, *Ceratium furca*, and *Glenodinium*.[68,69]

Since the availability of stored energy in the form of fixed carbon is necessary for cell division, the optimal timing of photosynthetic activation relative to the light/dark cycle and cell division may be critical. In many species in which carbon fixation, the light/dark cycle, and the division cycle are known, the phase relationships appear to be species specific. For example, in *Euglena gracilis*[70] and nitrogen-limited *Skeletonema costatum*[27] maximum carbon assimilation occurs 7 hr before cell division and the onset of darkness. In *Ditylum brightwellii*[43] and nitrogen-limited *Thalassiosira pseudonana*[44] carbon uptake is in phase with division; in *Ditylum brightwellii* both occur 1 to 2 hr before dark, and in *Thalassiosira pseudonana* both occur 6 hr before dark.

In *Ditylum brightwellii* the oscillation of total cell carbohydrate over the 24 hr light/dark cycle is correlated with the carbon assimilation rate.[43] The starch and lipid contents of *Scenedesmus*[63] and *Chlorella*[71,72] oscillate diurnally. In *Euglena gracilis* protein and several dialyzable subfractions of cellular protein also oscillate with circadian rhythmicity, while lipid phosphorus oscillates bimodally.[73]

It is often maintained that various macromolecular cell components are synthesized at a constant rate over the cell cycle.[29,74] This conclusion is usually based on the apparently continuous accumulation of cellular RNA, DNA, and protein during the cycle. These observations, however, can be misleading, as in the light/dark synchronized cell cycle of *Euglena gracilis*, several sharply defined, discontinuous peaks in DNA,[75,76] RNA,[77] and protein synthesis[73,78] have been found. Therefore, the biosynthesis of the macromolecules is not occurring at the same relative rates although their total accumulation may appear to be continuous.

VI. COEXISTENCE AND COMPETITIVE EXCLUSION

Two organisms without a periodic division cycle, but sharing a common limiting nutrient can coexist only when their growth rates are identical. Coexistence may also occur at the point where the Monod growth curves (Equation 9) for both species intersect.[12] These conditions can be expressed as

$$\mu_1 = \mu_2 = D = (V_1 \cdot X_1 + V_2 \cdot X_2)/(q_1 \cdot X_1 + q_2 \cdot X_2) \qquad (28)$$

and

$$S_c = \frac{K_{S_1}}{(\bar{\mu}_{m_1}/D) - 1} = \frac{K_{S_2}}{(\bar{\mu}_{m_2}/D) - 1} \qquad (29)$$

where S_c is the concentration at which the species can coexist and the numerical subscripts denote species 1 and 2. A slight change in S_c would ultimately result in competitive elimination of one species by the other. On the other hand, if S fluctuates around S_c in a relatively short time, they may coexist. These exacting conditions are probably not common and thus may not account for the diversity of organisms living together in the same natural environment.

In a natural environment two species may coexist if the growth of each is limited by different nutrients. Coexistence is possible because the growth rate of a single species is limited only by the nutrient in shorter supply, and other nutrients do not influence its growth.[21,38,39] Various organisms require nutrients in different proportions; for example, the optimum nitrogen/ phosphorus atomic ratio at which the limiting nutrient changes from one to the other varies widely from species to species: the ratio is 30 for *Scenedesmus* sp.,[38,39] 12 for *Asterionella formosa*,[79] and 21 for *Ankistrodesmus facaltus*.[79] Figure 2 shows that *Asterionella formosa* and *Scenedesmus* sp. can coexist in environments in which nitrogen/phosphorus atomic ratios range from 12 to 30 because *Asterionella formosa* is limited by phosphorus, but *Scenedesmus* sp. is limited by nitrogen. Below the ratio of 12 both species are nitrogen limited, but limitation is more severe in *Scenedesmus* sp. than in *Asterionella formosa*, and *Asterionella formosa* competitively eliminates *Scenedesmus* sp. Above a ratio of 30 both are limited by phosphorus; the competitive advantage is thus reversed and *Scenedesmus* sp. eliminates *Asterionella formosa*.

Coexistence in a cyclostat requires that period averages for μ (*t*) be equal,[18] which can be accomplished by temporal differences in phasing of division and nutrient uptake.[29,31,51] The conditions required for coexistence of two species in the cyclostat environment are that

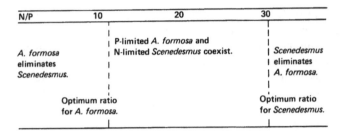

N/P	10	20	30
		P-limited *A. formosa* and N-limited *Scenedesmus* coexist.	
A. formosa eliminates *Scenedesmus.*			*Scenedesmus* eliminates *A. formosa.*
	Optimum ratio for *A. formosa.*		Optimum ratio for *Scenedesmus.*

FIGURE 2.   Predicted competitive interaction between *Asterionella formosa* and *Scenedesmus* sp. based on optimum N/P ratios. The optimum N/P ratio is defined as the ratio of subsistence quotas (viz. cell quota when $\mu = 0$) for N and P limited growth. See text for further details.

$$\langle \mu_1 \rangle_T = \langle \mu_2 \rangle_T = D = \langle \mu_1 \cdot X_1 \rangle_T / \langle X_1 \rangle_T = \langle \mu_2 X_2 \rangle_T / \langle X_2 \rangle_T = \frac{\langle V_1 \cdot X_1 \rangle_T + \langle V_2 \cdot X_2 \rangle_T}{\langle q_1 \cdot X_1 \rangle_T + \langle q_2 \cdot X_2 \rangle_T} \tag{30}$$

Thus in the periodic cyclostat environment, only the period averaged rather than the instantaneous values of these quantities need to be equal for coexistence.

Phase relationships of biosynthetic processes that culminate in division may vary considerably from species to species and from process to process. This different phasing effect of the light/dark cycle would increase the probability of coexistence because a single limiting nutrient may be temporally partitioned by two or more species. Thus competitive exclusion and coexistence based on nutrient ratios[12,39,80] are applicable only when competing organisms have identical phase relationships.

There are other theoretical treatments[18,31] for growth and coexistence in the literature, but it is not possible at present to predict coexistence in a cyclostat from the parameters of a single species.

Some differences in the timing of the uptake of limiting nutrients that may lead to coexistence are illustrated in Table 2. The maximum rate of nitrate and ammonium uptake in *Monochrysis lutheri*[81] is restricted to the subjective day, while the maximum for *Skeletonema costatum*[27] occurs during the subjective night. The rate of phosphorus uptake in *Euglena gracilis*,[28,30] *Asterionella formosa*,[82] and *Scenedesmus*[83] is maximal at various times during the subjective day. The rate of silicate uptake in *Skeletonema constatum*[84] is maximal approximately halfway through the subjective night, whereas for *Ditylum brightwellii*[84] the rate is maximal considerably earlier during the night.

Photosynthetic purple sulfur bacteria provide an excellent example of coexistence made possible by interspecies differences in the timing of the uptake of limiting nutrients during a photocycle. These bacteria utilize sulfide as the principal electron donor in photosynthesis. Van Gemerden[85] found that *Chromatium vinosum* competitively excluded *Chromatium wessei* at all growth rates in continuously illuminated, sulfide-limited chemostat cultures. However, these organisms coexist in their natural environment, where sulfide probably limits the growth of both species. They inhabit anaerobic aquatic environments exposed to a light/dark cycle. In a sulfide-limited chemostat culture exposed to a 6 hr/6 hr light/dark cycle, the two species soon achieved a stable coexistence. Neither species takes up sulfide during the dark and thus, sulfide accumulates during the dark. Coexistence occurred because *Chromatium wessei*, the slower growing organism of the two, could take up and oxidize the sulfide accumulated during the dark period at more than twice the rate as *Chromatium vinosum*. Sulfide is oxidized to molecular sulfur under illumination and partially stored as intracellular storage reserves, which is later utilized for growth. This phenomenon also illustrates the advantage of uncoupling nutrient uptake from division.

In natural assemblages of phytoplankton nutrient uptake is also periodic.[51,84,86-89] It is conceivable, therefore, that different nutrients may limit different species at different times of the day. Partitioning of nutrients on the basis of differences in the timing of maximum uptake and differences in critical extracellular and intracellular nutrient ratios[38,39,90] would minimize competitive overlap for a greater number of species and increase diversity in the ecosystem.

VII. TIMING OF CELL DIVISION

Daily oscillations in cell density in cyclostats contrast with the time-invariant cell densities in nonperiodic chemostats. Since cyclostats are relatively new and have been used to study only a few species, the following discussion includes information derived from studies of batch cultures, which will provide the basis for future work with cyclostats.

A variety of environmental factors appear to affect either the timing of cell division relative to the light/dark cycle (i.e., the phase angle) or the degree of phasing. Temperature, for example, appears to shift the division phase angle in some species. As shown in Figure 3, *Ditylum brightwellii* divides during the light period at 21°C, but at 15°C it divides mainly during the dark. Nelson and Brand[91] have noted similar, though less pronounced, phase shifts in *Thalassiosira pseudonana*. In some cases, temperature effects may explain the difference in the timing of division observed by various investigators. For example, Eppley et al.[43] and Paasche[93] observed that *Ditylum brightwellii* grown at 20°C divided during the light period, whereas Richmand and Rogers[94] and Smayda[6] found that the same species grown at near 15°C divided during the dark period. *Skeletonema costatum* also appears to shift division timing with change in temperature. At 18°C, this species divides during dark period,[27] but at 12°C it divides during the subjective day.[95] In this last case light intensity may also have influenced the phase shift. It was noted that at 20°C, the timing of division of *Skeletonema costatum* could be shifted from the dark period toward the light period simply by reducing intensity of the light.[96]

The effect of photoperiod itself on the timing of cell division is not clear. In some species, such as *Euglena gracilis*, division appears to start at a fixed time after the beginning of the light period regardless of its length, whereas in other species, such as *Dunaliella tertiolecta*, changes in photoperiod seem to cause phase shifts (Table 5). Although this phenomenon was examined in detail in *Ditylum brightwellii, Nitzschia turgidula*, and *Coccolithus huxleyi*[93,97] (Table 5), light intensity was varied simultaneously with photoperiod, and the effects of the two variables cannot be distinguished.

The effect of nutrient limitation on phased cell division is also not well understood. In large-scale outdoor continuous cultures of *Chaetoceros* sp.,[98] phased cell division was expressed only at high nutrient levels. However, in chemostat cultures of *Skeletonema costatum*,[27] *Coccolithus huxleyi*,[27] and *Euglena gracilis*[10,28] cell division is strongly phased even when nutrients are limited. There is also evidence that the degree of phasing in *Euglena gracilis* depends on growth rate or on the degree of nutrient limitation.[10]

It is difficult to generalize about the timing of cell division of various taxonomic groupings of phytoplankton. The data in Table 5 were obtained under a wide variety of experimental conditions that could obscure any systematic patterns that may exist. Perhaps the only general pattern is that the diatom species *(Bacillariophyceae)* tend to divide during the light period, while other taxonomic groups divide during the dark

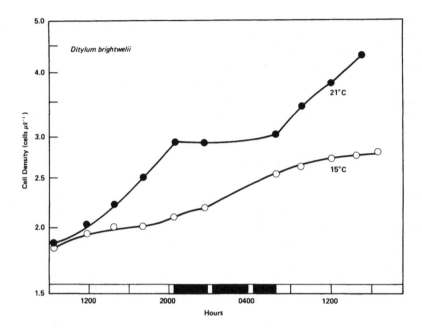

FIGURE 3.   Cell division patterns for *D. brightwellii* grown in batch culture on L:D
14/10 at 21°C and 15°C.[106]

period. It has been suggested that this difference could be associated with the interrelationships of silicification and cell division and the energy required for silicic acid uptake[91] — the latter an energy burden unique to the diatoms. Substantiation of this hypothesis needs conclusive demonstration that diatoms are truly daytime dividers, but exceptions are noted in Table 5.

The underlying mechanism for the phasing of cell division in *Gonyaulax polyedra*,[99] *Euglena gracilis*,[100] and *Chlamydomonas reinhardtii*[6] appears to be the coupling of the cell cycle to a biological clock. Because of this coupling, division is possible only when the clock is in a certain phase of its 24-hr cycle. Since the phase angle between the clock and the entraining light/dark cycle is the same for all cells in a population, they are forced to divide within a restricted interval during the light/dark cycle.

For other species, mechanisms that control the timing of cell division are unclear. For example, the cell division pattern of diatoms are not as consistent as those of clock-controlled species. Although diatoms possess endogenous rhythms in other processes,[101,102] there is no evidence that cell division is coupled to the clock. In this class, the timing of division appears to be particularly sensitive to environmental conditions, and it may be determined simply by the dynamics of the metabolic processes and their coupling to the daily supply of light energy.

VIII. CONCLUDING REMARKS

Time-independent physiological and biochemical mechanisms can be studied in non-periodic chemostats, but the results can be extrapolated to ecology to a relatively limited extent. As pointed out by Jannasch,[11] this experimental system was never intended to simulate natural environments. In particular, continuous illumination for photosynthetic organisms — or examination at fixed times during a light/dark cycle — obscures some important ecological attributes of an organism. In contrast, a cyclostat provides a diel light/dark cycle and thus yields ecological information such as differences in the

Table 5

Species	Reference	Phase Relationships	Species	Reference	Phase Relationships
	1			9	
	2		Chaetoceros sp.	10	
	2		Navicula ostrearia	11	
	2		Cachonia niei	12	
Ditylum brightwellii	2		Ceratium dens	13	
	3		Ceratium furca	13	
	4		Dinophysis fortii	13	
	5		Gonyaulax polyedra	14	
	6			15	
	7			15	
Skeletonema costatum	3		Dunaliella tertiolecta	15	
	7			15	
	8			15	
Phaeodactylum tricornutum	9		Euglena gracilis	16	
	2			16	
	2			17	
Nitzschia turgidula	2		Chlamydomonas reinhardtii	18	
	2		Chlamydomonas moewusii	19	
Thalassiosira fluviatilis	3				
Thalassiosira pseudonana	3		Chlorella fusca	20	
Chaetoceros gracilis	3		Scenedesmus obliquus	21	
Lithodesmium undulatum	3			22	
	9		Coccolithus huxleyi	22	
	9			22	
Chaetoceros armatum	9			22	

Note: Survey of cell division patterns in phytoplankton grown on light/dark cycles or natural populations *in situ*. Hatched areas represent the dark period and solid bars depict the period over which cell division occurs.

REFERENCES

1. **Eppley, R. W., Holmes, R. W., and Paasche, E.**, Periodicity in cell division and physiological behavior of *Ditylum brightwellii*, a marine plankton diatom, during growth in light-dark cycles, *Arch. Mikrobiol.*, 56, 305, 1967.

2. **Paasche, E.**, Marine plankton algae growth with light/dark cycles. II. *Ditylum brightwellii* and *Nitzschia turgidula*, *Physiol. Plant.*, 21, 66, 1968.

3. **Chisholm, S. W., Azam, R., and Eppley, R. W.**, Silicic acid incorporation in marine diatoms on light:dark cycles: use as an assay for phased cell division, *Limnol. Oceanogr.*, 23, 518, 1978.

4. **Smayda, T. J.**, Phased cell division in natural populations of the marine diatom *Ditylum brightwellii* and the potential significance of diel phytoplankton behavior in the sea, *Deep-Sea Res.*, 20, 151, 1975.

5. **Richman, S. and Rogers, J. N.**, The feeding of *Calanus helgolandicus* on synchronously growing populations of the marine diatom *Ditylum brightwellii*, *Limnol. Oceanogr.*, 14, 701, 1969.

6. **Eppley, R. W., Rogers, J. N., McCarthy, J. J., and Sournia, A.**, Light/dark periodicity in nitrogen assimilation of the marine phytoplankters *Skeletonema costatum* and *Coccolithus huxleyi* in N-limited chemostat culture, *J. Phycol.*, 7, 150, 1971.

7. **Jorgensen, E. G.**, Photosynthetic activity during the life cycle of synchronous *Skeletonema* cells, *Physiol. Plant.*, 19, 789, 1966.

8. **Harrison, P. J.**, Continuous Culture of the Marine Diatom *Skeletonema costatum* (Greve) Cleve, Under Silicate-Limitation, Ph.D. dissertation, University of Washington, Seattle, 1974.

9. **Palmer, J. D., Livingston, L., and Zusy, Fr. D.**, A persistant diurnal rhythm in photosynthetic capacity, *Nature, (London)*, 203, 1087, 1964.

10. **Malone, T. C., Garside, C., Haines, K. C., and Roels, O. A.**, Nitrate uptake and growth of *Chaetoceros* sp. in large outdoor continuous cultures., *Limnol. Oceanogr.*, 20, 9, 1975.

11. **Neuville, D. and Daste, P.**, Premiers essais de synchronisation des divisions cellulaires chez la Diatomée *Navicula ostrearia* (Gaillom) Bory en culture axénique, *C. R. Acad. Sci. Paris, Ser. D.*, 284, 761, 1977.

12. **Loeblich, A. R.**, Studies in synchronously dividing populations of *Cachonina niei*, a marine dinoflagellate, *Bull. Jap. Soc. Phycol.*, 25, (Suppl.), 119, 1977.

13. **Weiler, C. S. and Chisholm, S. W.**, Phased cell division in natural populations of marine dinoflagellates from shipboard cultures, *J. Exp. Mar. Biol. Ecol.*, 25, 239, 1976.

14. **Sweeney, B. M. and Hastings, J. W.**, Rhythmic cell division in populations of *Gonyaulax polyedra*, *J. Protozool.*, 5, 217, 1958.

15. **Eppley, R. W. and Coatsworth, J. L.**, Culture of the marine phytoplankter *Dunaliella tertiolecta* with light/dark cycles., *Arch. Mikrobiol.*, 55, 66, 1966.

16. **Edmunds, L. N.**, Studies on synchronously dividing cultures of *Euglena gracilis* Kelbs (Strain Z).I. Attainment and characterization of rhythmic cell division, *J. Cell. Comp. Physiol.*, 66, 147, 1965.

17. **Edmunds, L. N. and Funch, R.**, Effects of "skeleton" photoperiods and high frequency light/dark cycles on the rhythm of cell division in synchronized cultures of *Euglena*, *Planta (Berlin)*, 87, 134, 1969.

18. **Bruce, V. G.**, The biological clock in *Chlamydomonas reinhardtii*, *J. Protozool.*, 17, 328, 1970.

19. **Bernstein, E.**, Synchronous division in *Chlamydomonas moewusii*, *Science*, 131, 1528, 1960.

20. **Lorenzen, H.**, Synchronous cultures, in *Photobiology of Microorganisms*, Halldal, P., Ed., John Wiley & Sons Interscience New York, 1970, 187.

21. **Senger, H.**, Charakterisierung einer Synchronkultur von *Scenedesmus obliquus* ihrer potentiellen photosyntheseleistung und des Photosynthese-Quotienten während des Entwicklungscyclus., *Planta (Berlin)*, 90, 243, 1970.

22. **Paasche, E.**, Marine plankton algae grown with light/dark cycles. I. *Coccolithus huxleyi*, *Physiol. Plant.*, 20, 946, 1967.

timing of metabolic activities among various organisms, which cannot be obtained under continuous illumination.

The mathematical treatment of cyclostat growth kinetics is admittedly complex. However, we are only beginning to understand phased growth in cyclostat. Analogies between chemostat and cyclostat growth seem to indicate that with further investigation, easily manageable, basic, universal mathematical principles may gradually be developed. The next challenge in experimental studies of phytoplankton is to understand the effects of dynamic nutrient regimes. The understanding of growth and adaptive changes in physiology under these transient conditions will provide further insight into the ecology of phytoplankton.

IX. GLOSSARY OF VARIABLES AND CONSTANTS CITED IN TEXT

D = dilution rate (time^{-1})

DT = $_o\int^T \mu \cdot dt$, integrated growth over the period T (dimensionless)

D_{max}^T = maximum integrated growth over the period T (dimensionless)

i = intracellular pools of nutrient that regulate uptake of nutrients (μmole$\cdot$cell^{-1})

K_i = inhibition constant for uptake in terms of intracellular pools (μmole$\cdot$cell^{-1})

K_q = half-saturation constant for growth as a function of cell quota q (μmole$\cdot$cell^{-1})

K_{qc} = half-saturation constant for period averaged growth as a function of cell quota at the trough of its oscillation q_c (μmole$\cdot$cell^{-1}). This is the cell quota when $<u>_T = <u_m>_T/2$.

K_m = half-saturation constant for nutrient uptake (μM)

K_s = half-saturation constant for growth as a function of extracellular nutrient concentrations (μM)

$\bar{K}_s$ = half-saturation constant for integrated growth as a function of S. This is the concentration when $(e^{DT} - 1) = (e^{D_{max} \cdot T} - 1)/2$.

K_{so} = half-saturation constant for period averaged growth as a function of S_o (μM)

q = cell quota, or the amount of intracellular limiting nutrient (μmole$\cdot$cell^{-1})

q_c = minimum cell quota for the period T, or the level of intracellular nutrient at the trough of its oscillation (μmole$\cdot$cell^{-1})

q_o = subsistence cell quota, or the value of intracellular nutrient concentrations when $\mu = 0$ (μmole$\cdot$cell^{-1}).

q_{co} = subsistence cell quota in terms of q_c, or the value of q_c when $<\mu>_T = 0$ (μmole$\cdot$cell^{-1}).

q_m = maximum cell quota, or the value of cell quota when it is independent of S or μ (μmole$\cdot$cell^{-1}); for example, the cell quota of phosphorus in nitrogen-limited cells is q_m for phosphorus.

S = extracellular nutrient concentration (μM)

$\bar{S}$ = average extracellular nutrient concentration for the period T (μM)

S_c = extracellular concentration of nutrient at which two competing species can coexist (μM)

S_o = extracellular nutrient concentration at the peak of its oscillation (μM)

S_r = concentration of limiting nutrient in the reservoir (μM)

T = period length (24 hr) for oscillations

t	=	time during the period T
V	=	uptake rate of nutrient (μmole·cell^{-1}·min^{-1})
V_m	=	true maximum uptake velocity for nutrient (μmole·cell^{-1}·min^{-1}), or the rate of uptake when S is at saturation levels and i or μ is equal to 0.
$\hat{V}_m$	=	apparent maximum uptake velocity of nutrient (μmole·cell^{-1}·min^{-1}), or the rate of uptake for a cell with given amount of i when S is saturating.
X	=	concentration of cells or biomass (cells·mℓ^{-1})
X_{max}	=	concentration of cells at the peak of its oscillation (cells·mℓ^{-1})
X_{min}	=	concentration of cells at the trough of its oscillation (cells·mℓ^{-1})
Z	=	$(1 + i/K_i)$ inhibition term for nutrient uptake (dimensionless)
μ	=	specific growth rate (time^{-1})
$\bar{\mu}_m$	=	theoretical maximum specific growth rate is expressed as a function of external nutrient concentrations, or the specific growth rate when S is infinite (time^{-1}).
$\hat{\mu}_m$	=	theoretical maximum specific growth rate when growth rate is expressed as a function of cell quota, or the growth rate when q is infinite (time^{-1}).
μ_m	=	true physiological maximum growth rate for nutrient unlimited growth (time^{-1}).

ACKNOWLEDGMENTS

This work has been supported in part by EPA grant #R 804689 and NSF grant #DEB 75-19519 to G-Y.R., and NSF grant #OCE 77-08999 to S.W.C. The authors would also like to acknowledge J. A. Blanckaert, B. Kusel, and M. E. Rudolph for their assistance in the preparation of this manuscript.

REFERENCES

1. **Hastings, J. W. and Sweeney, B. M.,** Phased cell division in the marine dinoflagellates, in *Synchrony in Cell Division and Growth,* Zeuthen, E., Ed., John Wiley & Sons, Interscience, New York, 1964, 307.
2. **Lorenzen, H.,** Synchronous cultures, in *Photobiology of Microorganisms* , Halldal, P., Ed., J. Wiley & Sons, Interscience, New York, 1970, 187.
3. **Lorenzen, H. and Heese, M.,** Synchronous cultures, in *Algal Physiology and Biochemistry,* Stewart, W. D. P., Ed., Bot. Monogr., Vol. 10, Blackwell Scientific Publications, Oxford, 1974, 894.
4. **Pirson, A. and Lorenzen, H.,** Synchronized dividing algae, *Annu. Rev. Plant Physiol.,* 17, 439, 1966.
5. **Tamiya, H.,** Synchronous cultures of algae, *Annu. Rev. Plant Physiol.,* 17, 1, 1966.
6. **Smayda, T.,** Phased cell division in natural populations of the marine diatom *Ditylum brightwellii* and the potential significance of diel phytoplankton behavior in the sea, *Deep-Sea Res.,* 22, 151, 1975.
7. **Sournia, A.,** Circadian periodicities in natural populations of marine phytoplankton, *Adv. Mar. Biol.,* 12, 325, 1974.
8. **Swift, E. and Durdin, E. G.,** The phased cell division and cytological characteristics of *Pyrocystis* spp. can be used to estimate doubling times of their populations in the sea, *Deep-Sea Res.,* 19, 189, 1972.

9. **Weiler, C. S. and Chisholm, S. W.**, Phased cell division in natural population of marine dinoflagellates from shipboard cultures, *J. Exp. Mar. Biol. Ecol.*, 25, 239, 1976.

10. **Chisholm, S. W., Stross, R. G., and Nobbs, P. A.**, Light/dark-phased cell division in *Euglena gracilis*(Z) (Euglenophyceae) in PO_4-limited continuous culture, *J. Phycol.*, 11, 367, 1975.

11. **Jannasch, H. W.**, Steady state and the chemostat in ecology, *Limnol. Oceanogr.*, 19, 716, 1974.

12. **Rhee, G-Y.**, Continuous culture in phytoplankton ecology, in *Advances in Aquatic Microbiology*, Vol. 2, Droop, M. R. and Jannasch, H. W., Eds., Academic Press, London, 1980, 151—203.

13. **Halvorson, H. O., Carter, B. L. A., and Tauro, P.**, Synthesis of enzymes during the cell cycle, in *Advances in Microbial Physiology*, Vol. 6, Rose, A. H. and Wilkinson, J. F., Eds., Academic Press, New York, 1971, 47.

14. **Howell, S. H.**, An analysis of cell cycle controls in temperature-sensitive mutants of *Chlamydomonas reinhardtii*, in *Cell Cycle Controls*, Padilla, G. M., Cameron, I. L., and Zimmerman, A. Eds., Academic Press, New York, 1974, 235.

15. **Mitchison, J. M.**, *The Biology of the Cell Cycle*, Cambridge University Press, Cambridge, 1971.

16. **Mitchison, J. M.** Sequences, pathways, and timers in the cell cycle, in *Cell Cycle Controls*, Padilla, G. M., Cameron, I. L., and Zimmerman, A., Eds., Academic Press, New York, 1974, 125.

17. **Frisch, H. L. and Gotham, I. J.**, On periodic algal cyclostat populations, *J. Theor. Biol.*, 66, 665, 1977.

18. **Frisch, H. L. and Gotham, I. J.**, A simple model for periodic cyclostat growth of algea, *J. Math. Biol.*, 7, 149, 1979.

19. **Monod, J.**, La technique et culture continue: théorie et applications, *Ann. Inst. Pasteur (Paris)*, 79, 390, 1950.

20. **Droop, M. R.**, Vitamin B_{12} and marine ecology. IV. The kinetics of uptake, growth, and inhibition in *Monochrysis lutheri*, *J. Mar. Biol. Assoc. U.K.*, 48, 689, 1968.

21. **Droop, M. R.**, The nutrient status of algal cells in continuous culture, *J. Mar. Biol. Assoc. U.K.*, 54, 825, 1974.

22. **Tempest, D. W.**, The continuous cultivation of micro-organisms, in *Methods in Microbiology*, Vol. 2, Norris, J. R. and Ribbons, D. W., Eds., Academic Press, London, 1969, 259.

23. **Caperon, J. and Meyer, J.**, Nitrogen-limited growth of marine phytoplankton. I. Changes in population characteristics with steady-state growth rate, *Deep-Sea Res.*, 19, 601, 1972.

24. **Müller, H.**, Wachstum und Phosphatebedarf von *Nitzschi actinastroides* (Lemm.) v. Goor in statischer und homokontinuierlicher Kulter unter Phosphatlimitierung, *Arch. Hydrobiol.*, *Suppl.*, 38, 399, 1972.

25. **Paasche, E.**, Silicon and the ecology of marine diatoms. I., *Thalassiosira pseudonana* Hasle and Heimdal (*Cyclotella nana* Hustedt) grown in a chemostat with silicate as limiting nutrient, *Mar. Biol.*, 19, 117, 1973.

26. **Bruce, V. G.**, The biological clock in *Chlamydomonas reinhardtii*, *J. Protozool.*, 17, 328, 1970.

27. **Eppley, R. W., Rogers, J. N., McCarthy, J. J., and Sournia, A.**, Light/dark periodicity in nitrogen assimilation of the marine phytoplankters *Skeletonema costatum* and *Coccolithus huxleyi* in N-limited chemostat culture, *J. Phycol.*, 7, 150, 1971.

28. **Gotham, I. J.**, Nutrient Limited Cyclostat Growth: A Theoretical and Physiological Analysis, Ph.D. thesis, State University of New York, Albany, 1977.

29. **Williams, F. M.**, Dynamics of microbial populations, in *Systems Analysis and Simulation in Ecology*, Patten, B. C., Ed., Academic Press, New York, 1971, 197.

30. **Chisholm, S. W. and Stross, R. G.**, Phosphate uptake kinetics in *Euglena gracilis*(Z) (Euglenophyceae) grown in light/dark cycles. II. Phased PO_4-limited culture, *J. Phycol.*, 12, 217, 1976.

31. **Chisholm, S. W. and Nobbs, P. A.**, Simulation of algal growth and competition in a phosphate-limited cyclostat, in *Modeling Biochemical Processes in Aquatic Ecosystems*, Canale, R. P., Ed., Ann Arbor Science Publishers, Ann Arbor, 1975, 337.

32. **Caperon, J.**, Population growth response of *Isochrysis galbana* to nitrate variation at limiting concentrations, *Ecology*, 49, 867, 1969.

33. **Fuhs, G. W.**, Phosphorus content and rate of growth in the diatoms *Cyclotella nana* and *Thalassiosira fluviatilis*, *J. Phycol.*, 5, 312, 1968.

34. **Droop, M. R.**, The nutrient status of algal cells in batch culture, *J. Mar. Biol. Assoc., U.K.*, 55, 541, 1975.

35. **Tilman, D. and Kilham, S. S.**, Phosphate and silicate growth and uptake kinetics of the diatoms *Asterionella formosa* and *Cyclotella menqhiniana* in batch and semicontinuous culture, *J. Phycol.*, 12, 375, 1976.

36. **Tett, P., Cottrell, J. C., Trew, D. O., and Wood, B. J. B.**, Phosphorus quota and the chlorophyll carbon ratio in marine phytoplankton, *Limnol. Oceanogr.*, 20, 587, 1975.

37. **Barlow, J. P.**, Cell quotas of phosphorus and silicon and growth of diatoms in Cayuga Lake, *Limnol. Oceanogr.*, in press, 1980.

38. **Rhee, G-Y.**, Phosphate uptake under nitrate limitation by *Scenedesmus* sp. and its ecological implications, *J. Phycol,.* 10, 470, 1974.
39. **Rhee, G-Y.**, Effects of N:P atomic ratios and nitrate limitation on algal growth, cell composition, and nitrate uptake, *Limnol. Oceanogr.*, 23, 10, 1978.
40. **Droop, M. R.**, Nutrient limitation in osmotrophic protista, *Am. Zool.*, 13, 209, 1973.
41. **Droop, M. R.**, Some thoughts on nutrient limitation in algae, *J. Phycol,.* 9, 264, 1973.
42. **Droop, M. R.**, Vitamin B_{12} and marine ecology. V. Continuous culture as an approach to nutritional kinetics, *Helgol. Wiss. Meeresunters.*, 20, 629, 1970.
43. **Eppley, R. W., Holmes, R. W., and Paasche, E.**, Periodicity in cell division and physiological behavior of *Ditylum brightwellii*, a marine planktonic diatom, during growth in light-dark cycles, *Arch. Mikrobiol.*, 56, 305, 1967.
44. **Eppley, R. W. and Renger, E. H.**, Nitrogen assimilation of an oceanic diatom in nitrogen-limited continuous culture, *J. Phycol.*, 10, 15, 1974.
45. **Rhee, G-Y.**, A continuous culture study of phosphate uptake, growth, and polyphosphate in *Scenedesmus* sp., *J. Phycol.*, 9, 495, 1973.
46. **Perry, M. J.**, Phosphate utilization by an oceanic diatom in phosphorus-limited chemostat culture and in the oligotrophic waters of the central North Pacific, *Limnol. Oceanogr.*, 21, 88, 1976.
47. **Gotham, I. J. and Rhee, G-Y.**, Comparative kinetic studies of phosphate limited growth and phosphate uptake in phytoplankton in continuous culture, in preparation.
48. **Guillard, R. R. L., Kilham, P., and Jackson, T. A.**, Kinetics of silicon-limited growth in the marine diatom *Thalassiosira pseudonana* Hasle and Heimdal (= Cyclotella nana Hustedt), *J. Phycol.*, 9, 233, 1975.
49. **Gotham, I. J. and Rhee, G-Y.**, Comparative kinetic studies of nitrate limited growth and nitrate uptake in phytoplankton in continuous culture, in preparation.
50. **Chisholm, S. W. and Stross, R. G.**, Phosphate uptake kinetics in *Euglena gracilis* (Z) (Euglenophyceae) grown on light/dark cycles. I. Synchronized batch cultures, *J. Phycol.*, 12, 210, 1976.
51. **Stross, R. G. and Pemrick, S. M.**, Nutrient uptake kinetics in phytoplankton: a basis for niche separation, *J. Phycol.*, 10, 164, 1974.
52. **Edmunds, L. N.**, Circadian clock control of the cell developmental cycle in synchronized cultures of *Euglena*, in Mechanisms of Regulation of Plant Growth, *R. Soc. N. Z., Bull.*, 12, 287, 1974.
53. **Edmunds, L. N.**, On the interplay among cell cycle, biological clock, and membrane transport control systems, *Int. J. Chronobiol.*, 2, 233, 1974.
54. **Ehret, C. F.**, The sense of time: evidence for its molecular basis in the eucaryotic gene action system, *Adv. Biol. Med. Phys.*, 15, 17, 1974.
55. **Ehret, C. F., Meinert, J. C., Groh, K. R., Dobra, K. W., and Antipa, G. R.**, Circadian regulation: growth kinetics of the infradian cell, in *Growth Kinetics and Biochemical Regulation of Normal and Malignant Cells*, A Collection of Papers Presented at the 29th Annu. Symp. Fundamental Cancer Res., Williams & Wilkins, Baltimore, 1977, 49—76.
56. **Baker, A. L. and Schmidt, R. R.**, Induced utilization of polyphosphate during nuclear division in synchronously growing *Chlorella, Biochim. Biophys. Acta*, 93, 180, 1964.
57. **Miyachi, S., Kanai, R., Mihara, S., Miyachi, S., and Aoki, S.**, Metabolic roles of inorganic polyphosphate in *Chlorella* cells, *Biochim. Biophys. Acta*, 93, 625, 1964.
58. **Ullrich, W. R.**, Der einflub von CO_2 und pH auf die ^{32}P-markierung von polyphosphaten und organischen phosphaten bei *Ankistrodesmus braunii* im licht, *Planta (Berlin)*, 102, 37, 1972.
59. **Ullrich, W. R.**, Untersuchungen über die raten der polyphosphatsynthese durch die photophosphorylierung die *Ankistrodesmus braunii, Arch. Mikrobiol.*, 87, 323, 1972.
60. **Herrmann, E. C. and Schmidt, R. R.**, Synthesis of phosphorus-containing macromolecules during synchronous growth of *Chlorella pyrenoidosa, Biochim. Biophys. Acta*, 95, 63, 1965.
61. **Sicko-Goad, L. and Stoermer, E. F.**, A morphometric study of lead and copper effects on *Diatoma tenue* var. *elongatum* (Bacillariophyta), *J. Phycol.*, 15, 316, 1979.
62. **Domanski-Kaden, J. and Simonis, W.**, Veränderungen der phosphat Fraktionen, besonders des Polyphosphates, bei synchronisierten *Ankistrodesmus brunni*-Kulturen, *Arch. Mikrobiol.*, 87, 11, 1972.
63. **Sundberg, I. and Nilshammar-Holmvall, M.**, The diurnal variation in phosphate uptake and ATP level in relation to deposition of starch, lipid, and polyphosphate in synchronized cells of *Scenedesmus, Z. Pflanzenphysiol.*, 76, 270, 1975.
64. **Cohen, D. and Parnas, H.**, An optimal policy for the metabolism of storage materials in unicellular algae, *J. Theor. Biol.*, 56, 1, 1976.
65. **Parnas, H. and Cohen, D.** The optimal strategy for the metabolism of reserve materials in microorganisms, *J. Theor. Biol.*, 56, 19, 1976.

66. **Eppley, R. W. and Coatsworth, J. L.**, Culture of the marine phytoplankter, *Dunaliella tertiolecta*, with light-dark cycles, *Arch. Mikrobiol.*, 55, 66, 1966.
67. **Myers, J. and Grahm, J. R.**, Photosynthetic unit size during the synchronous life cycle of *Scenedesmus, Plant Physiol.*, 55, 686, 1975.
68. **Prezélin, B. B. and Sweeney, B. M.**, Characterization of photosynthetic rhythms in marine dinoflagellates. II. Photosynthesis — irradiance curves and *in vivo* chlorophyll *a* fluorescence, *Plant Physiol.*, 60, 388, 1977.
69. **Prezélin, B. B., Meeson, B. W., and Sweeney, B. M.**, Characterization of photosynthetic rhythms in marine dinoflagellates. I. Pigmentation, photosynthetic capacity, and respiration, *Plant Physiol.*, 60, 384, 1977.
70. **Walther, W. G. and Edmunds, L. N.**, Studies on the control of the rhythm of photosynthetic capacity in synchronized cultures of *Euglena gracilis* (Z), *Plant Physiol.*, 51, 250, 1973.
71. **Wanka, F., Joppen, M. M. J., and Kuyper, CH. M. A.**, Starch-degrading enzymes of *Chlorella, Z. Pflanzenphysiol. Bd.*, 62, 146, 1970.
72. **Atkinson, A. W., Jr., John, P. C. L., and Gunning, B. E. S.**, The growth and division of the sengie mitochondrion and other organelles during the cell cycle of *Chlorella*, studied by quantitative sterology and three dimensional reconstruction, *Protoplasma*, 81, 77, 1974.
73. **Edmunds, L. N.**, Studies on synchronously dividing cultures of *Euglena gracilis* Klebs (Strain Z). II. Patterns of biosynthesis during the cell cycle, *J. Cell Comp. Physiol.*, 66, 159, 1965.
74. **Schmidt, R. B.**, Transcriptional and post-transcriptional control of enzyme levels in eucaryotic microorganisms, in *Cell Cycle Controls.*, Padilla, G. M., Cameron, I. L., and Zimmerman, A., Eds., Academic Press, New York, 1974, 201.
75. **Cook, J. R.**, Studies on chloroplast replication in synchronized *Euglena*, in *Cell Synchrony*, Cameron, I. L. and Padilla, G. M., Eds., Academic Press, New York, 1966, 153.
76. **Brandt, P.**, Zwei maxima plastidarer DNA-synthese mit unterschiedlicher lichtabha ngigkeit im zellcyclus von *Euglena gracilis, Planta (Berlin)*, 124, 105, 1975.
77. **Ledoigt. G., Calvayrac, R., Orcival-Lafont, A. M., and Pineau, B.**, Comparaison, à la lumiere et à l'obscurité, des synthèses specifiques d'ARN chez *Euglena gracilis* en cultures synchrones sur milieu lactate et étude du chrondriome, *Planta (Berlin)*, 103, 254, 1972.
78. **Feldman, J. F.**, Circadian rhythmicity in amino acid incorporation in *Euglena gracilis, Science*, 160, 1454, 1968.
79. **Rhee, G-Y. and Gotham, I. J.**, Optimum N:P ratios and coexistence of planktonic algae, *J. Phycol.*, in press, 1980.
80. **Tilman, D.**, Resource competition between planktonic algae: an experimental and theoretical approach, *Ecology*, 58, 338, 1977.
81. **Caperon, J. and Ziemann, D. A.**, Synergistic effects of nitrate and ammonium ion on the growth and uptake kinetics of *Monochrysis lutheri* in continuous culture, *Mar. Biol.*, 36, 73, 1976.
82. **Gotham, I. J. and Rhee, G-Y.**, in preparation, 1980.
83. **Azad, H. S. and Borchardt, J. A.**, Variations in phosphate uptake by algae, *Environ. Sci. Technol.*, 4, 737, 1970.
84. **Chisholm, S. W., Azam, F., and Eppley, R. W.**, Silicic acid incorporation in marine diatoms on light: dark cycles: use as an assay for phased cell division, *Limnol. Oceanogr.*, 23, 518, 1978.
85. **Van Gemerden, H.**, Coexistence of organisms competing for the same substrate: an example among the purple sulfur bacteria, *Microbial Ecol.*, 1, 104, 1974.
86. **MacIsaac, J. J.**, Diel cycles of inorganic nitrogen uptake in a natural phytoplankton population dominated by *Gonyaulax polyedra, Limnol. Oceanogr.*, 23, 1, 1978.
87. **Goering, J. J., Dugdale, R. C., and Menzel, D. W.**, Cyclic diurnal variations in the uptake of ammonia and nitrate by photosynthetic organisms in the Sargasso Sea, *Limnol. Oceanogr.*, 9, 448, 1964.
88. **Eppley, R. W., Carlucci, A. F., Holm-Hansen, O., Kiefer, D., McCarthy, J. J., Venrich, E., and Williams, P. M.**, Phytoplankton growth and composition in shipboard cultures supplied with nitrate, ammonium, or urea as the N-source, *Limnol. Oceanogr.*, 16, 741, 1971.
89. **Eppley, R. W., Packard, T. T., and MacIsaac, J. J.**, Nitrate reductase in Peru current phytoplankton, *Mar. Biol.*, 6, 195, 1970.
90. **Titman, D.**, Ecological competition between algae: experimental confirmation of resource-based competition theory, *Science*, 192, 463, 1976.
91. **Nelson, D. M. and Brand, L. E.**, Periodicity of cell division in 26 planktonic marine algae grown on light/dark cycles, submitted to *J. Phycol.*, 15, 67, 1979.
92. **Lewin, J. C., Reinmann, B. E., Busby, W. F., and Volcani, B. E.**, Silica shell formation in synchronously dividing diatoms, in *Cell Synchrony*, Cameron, I. L. and Padilla, G. M., Eds., Academic Press, New York, 1966, 160.
93. **Paasche, E.**, Marine plankton algae grown with light/dark cycles. II. *Ditylum brightwellii* and *Nitzschia turgidula, Physiol. Plant.*, 21, 66, 1968.

94. Richman, S. and Rogers, J. N., The feeding of *Calanus helgolandicus* on synchronously growing populations of the marine diatom *Ditylum brightwellii, Limnol. Oceanogr.,* 14, 701, 1969.

95. Harrison, P. J., Continuous Culture of the Marine Diatom *Skeletonema costatum* (Greve) Cleve, Under Silicate Limitation, Ph.D. thesis, University of Washington, Seattle, 1974.

96. Jorgensen, E. G., Photosynthetic activity during the life-cycle of synchronous *Skeletonema* cells, *Physiol. Plant.,* 19, 789, 1966.

97. Paasche, E., Marine plankton algae grown in light/dark cycles. I. *Coccolithus huxleyi, Physiol. Plant.,* 20, 946, 1967.

98. Malone, T. C., Garside, C., Haines, K. C., and Roels, O. A., Nitrate uptake and growth of *Chaetoceros* sp. in large outdoor continuous cultures, *Limnol. Oceanogr.,* 20, 9, 1975.

99. Sweeney, B. M. and Hastings, J. W., Rhythmic cell division in populations of *Gonyaulax polyedra, J. Protozool.,* 5, 217, 1958.

100. Edmunds, L. N., Studies on synchronously dividing cultures of *Euglena gracilis* Klebs (strain Z). III. Circadian components of cell division, *J. Cell Physiol.,* 67, 35, 1966.

101. Palmer, J. D. and Round, F. E., Persistant vertical migration rhythms in the benthic microflora. I. The effect of light and temperature on the rhythmic behavior of *Euglena obtusa, J. Mar. Biol. Assoc. U.K.,* 45, 567, 1967.

102. Fischer, H., Groning, C., and Koster, C., Vertical migration rhythm in freshwater diatoms, *Hydrobiology,* 56, 259, 1977.

103. Sicko-Goad, L., personal communication.

104. Frisch, H. L. and Gotham, I. J., A simple model for periodic cyclostat growth of algae, *J. Math. Biol.,* 7, 149, 1979.

105. Gotham, I. J., *Nutrient Limited Cyclostat Growth; a Theoretical and Physiological Analysis,* Ph. D. thesis, State University of New York, Albany, 1977.

106. Chisholm, S. W. and Karl, D., unpublished data, 1978.

INDEX

A

T - #0658 - 101024 - C0 - 253/174/12 - PB - 9781138558267 - Gloss Lamination